Transparency
in Statistical Information
for the National Center for Science and Engineering Statistics and All Federal Statistical Agencies

Panel on Transparency and Reproducibility of Federal Statistics for the National Center for Science and Engineering Statistics

Committee on National Statistics

Division of Behavioral and Social Sciences and Education

A Consensus Study Report of

The National Academies of
SCIENCES • ENGINEERING • MEDICINE

THE NATIONAL ACADEMIES PRESS
Washington, DC
www.nap.edu

THE NATIONAL ACADEMIES PRESS   500 Fifth Street, NW   Washington, DC 20001

This activity was supported by a contract between the National Academies of Sciences, Engineering, and Medicine and the National Science Foundation under grant number 1822391. Any opinions, findings, conclusions, or recommendations expressed in this publication do not necessarily reflect the views of any organization or agency that provided support for the project.

International Standard Book Number-13: 978-0-309-27045-8
International Standard Book Number-10: 0-309-27045-6
Digital Object Identifier: https://doi.org/10.17226/26360

Additional copies of this publication are available from the National Academies Press, 500 Fifth Street, NW, Keck 360, Washington, DC 20001; (800) 624-6242 or (202) 334-3313; http://www.nap.edu.

Copyright 2022 by the National Academy of Sciences. All rights reserved.

Printed in the United States of America

Suggested citation: National Academies of Sciences, Engineering, and Medicine. 2022. *Transparency in Statistical Information for the National Center for Science and Engineering Statistics and All Federal Statistical Agencies*. Washington, DC: The National Academies Press. https://doi.org/10.17226/26360.

The National Academies of
SCIENCES • ENGINEERING • MEDICINE

The **National Academy of Sciences** was established in 1863 by an Act of Congress, signed by President Lincoln, as a private, nongovernmental institution to advise the nation on issues related to science and technology. Members are elected by their peers for outstanding contributions to research. Dr. Marcia McNutt is president.

The **National Academy of Engineering** was established in 1964 under the charter of the National Academy of Sciences to bring the practices of engineering to advising the nation. Members are elected by their peers for extraordinary contributions to engineering. Dr. John L. Anderson is president.

The **National Academy of Medicine** (formerly the Institute of Medicine) was established in 1970 under the charter of the National Academy of Sciences to advise the nation on medical and health issues. Members are elected by their peers for distinguished contributions to medicine and health. Dr. Victor J. Dzau is president.

The three Academies work together as the **National Academies of Sciences, Engineering, and Medicine** to provide independent, objective analysis and advice to the nation and conduct other activities to solve complex problems and inform public policy decisions. The National Academies also encourage education and research, recognize outstanding contributions to knowledge, and increase public understanding in matters of science, engineering, and medicine.

Learn more about the National Academies of Sciences, Engineering, and Medicine at **www.nationalacademies.org**.

The National Academies of
SCIENCES · ENGINEERING · MEDICINE

Consensus Study Reports published by the National Academies of Sciences, Engineering, and Medicine document the evidence-based consensus on the study's statement of task by an authoring committee of experts. Reports typically include findings, conclusions, and recommendations based on information gathered by the committee and the committee's deliberations. Each report has been subjected to a rigorous and independent peer-review process and it represents the position of the National Academies on the statement of task.

Proceedings published by the National Academies of Sciences, Engineering, and Medicine chronicle the presentations and discussions at a workshop, symposium, or other event convened by the National Academies. The statements and opinions contained in proceedings are those of the participants and are not endorsed by other participants, the planning committee, or the National Academies.

For information about other products and activities of the National Academies, please visit www.nationalacademies.org/about/whatwedo.

PANEL ON TRANSPARENCY AND REPRODUCIBILITY OF FEDERAL STATISTICS FOR THE NATIONAL CENTER FOR SCIENCE AND ENGINEERING STATISTICS

DANIEL KASPRZYK (*Chair*), NORC at the University of Chicago
PHILIP ASHLOCK, GSA Technology Transformation Services, General Services Administration
DAVID BARRACLOUGH, Practices and Solutions Division, Organisation for Economic Co-operation and Development
CHRISTOPHER CHAPMAN, Sample Surveys Division, National Center for Education Statistics
DANIEL W. GILLMAN, Office of Survey Methods Research, U.S. Bureau of Labor Statistics
LINDA A. JACOBSEN, Population Reference Bureau, Inc.
H. V. JAGADISH, Department of Computer Science and Engineering, University of Michigan
FRAUKE KREUTER, Joint Program in Survey Methodology, University of Maryland
MARGARET LEVENSTEIN, Inter-university Consortium for Political and Social Research, University of Michigan
PETER V. MILLER, U.S. Census Bureau (retired)
AUDRIS MOCKUS, Department of Electrical Engineering and Computer Science, University of Tennessee, Knoxville
SARAH M. NUSSER, Center for Survey Statistics and Methodology, Iowa State University
ERIC RANCOURT, Modern Statistical Methods and Data Science Branch, Statistics Canada
WILLIAM L. SCHERLIS,* School of Computer Science, Carnegie Mellon University
LARS VILHUBER, Department of Economics, Cornell University

*Resigned from panel on October 28, 2019

MICHAEL L. COHEN, *Senior Program Officer*
MICHAEL SIRI, *Associate Program Officer*
CONNIE F. CITRO, *Senior Scholar*
JILLIAN KAUFMAN, *Program Coordinator* (until January 15, 2020)
ANTHONY MANN, *Program Coordinator*
JOHN GAWALT, *Consultant* (until May 18, 2020)

COMMITTEE ON NATIONAL STATISTICS

ROBERT M. GROVES (*Chair*), Office of the Provost, Department of Mathematics and Statistics and Department of Sociology, Georgetown University
LAWRENCE D. BOBO, Department of Sociology, Harvard University
ANNE C. CASE, Woodrow Wilson School of Public and International Affairs, Princeton University
MICK P. COUPER, Survey Research Center, Institute for Social Research, University of Michigan
JANET M. CURRIE, Woodrow Wilson School of Public and International Affairs, Princeton University
DIANA FARRELL, JPMorgan Chase Institute, Washington, DC
ROBERT GOERGE, Chapin Hall at The University of Chicago
ERICA L. GROSHEN, The ILR School, Cornell University
HILARY HOYNES, Goldman School of Public Policy, University of California, Berkeley
DANIEL KIFER, Department of Computer Science and Engineering, The Pennsylvania State University
SHARON LOHR, Consultant and Freelance Writer
JEROME P. REITER, Department of Statistical Science, Duke University
JUDITH A. SELTZER, Department of Sociology, University of California, Los Angeles
C. MATTHEW SNIPP, Department of Sociology, Stanford University
ELIZABETH A. STUART, Department of Mental Health, Johns Hopkins Bloomberg School of Public Health
JEANETTE WING, Data Science Institute, Columbia University

BRIAN HARRIS-KOJETIN, *Board Director*
MELISSA CHIU, *Deputy Board Director*
CONNIE F. CITRO, *Senior Scholar*

Acknowledgments

A Consensus Study Panel requires many individuals to assist the panel in studying the issues identified in the panel's statement of task. The Panel on Transparency and Reproducibility of Federal Statistics for the National Center for Science and Engineering Statistics is no different. Many experts were called upon to discuss issues, provide their expertise, and discuss their perspectives for the panel's consideration. The panel thanks all these individuals for the assistance and knowledge.

The panel benefitted greatly from the presentations provided in its open sessions. The experts the panel heard from can be clustered into the following perspectives and areas of expertise (see Appendix C for the agendas for open meetings): NCSES staff: Emilda Rivers, May Aydin, Tiffany Julian, and Francisco Moris; experts in metadata standards as used internationally: Olivier Dupriez (World Bank), Pascal Heus (Metadata Technology North America), Heidi Koumarianos (Institut National de la Statistique et des Études Économiques), and Juan Munoz (National Institute of Statistics and Geography, Mexico); experts from the federal statistical system: William Bell (Census Bureau), Marcus Berzofsky (RTI International), Christopher Carrino (Census Bureau), Leighton L Christiansen (Bureau of Transportation Statistics), Brad Edwards (Westat), John Eltinge (Census Bureau), Dennis Fixler (Bureau of Economic Analysis), Nick Hart (Data Coalition), Nancy Potok (formerly Office of Management and Budget), Mark Prell (Economic Research Service), Marilyn Seastrom (National Center for Education Statistics), Tori Velkoff (Census Bureau), and Zack Whitman (Census Bureau); experts in computer science: Jeremy Iverson and Dan Smith (Colectica), and Natasha Noy (Google); experts in administrative records data: John Czajka and Mathew Stange (Mathematica Policy Research); and an expert in the federal statistical user community: Jason Jurjevich (University of Arizona). We also heard from expert users of NCSES data: Kimberlee Eberle-Sudre (Association of American Universities) and Anne-Marie Knott (Washington University in St. Louis).

In addition to these public presentations, panel and staff participated in meetings and conference calls with staff from NCSES and the Interagency Council on Statistical Policy as well as George Alter (Inter-university Consortium for Political and Social Research), Jeremy Iverson (Colectica), and Rolf Schmitt and Leighton L Christiansen (Bureau of Transportation Statistics). Further, to gain insight into what is currently carried out in major statistical programs in terms of documentation and archival policy, the panel sent an informal questionnaire to the leaders of 20 programs of the federal statistical system, receiving responses from 11. The results of this questionnaire are provided in Chapter 2.

The panel and staff also studied a number of domestic and international documents that called for greater openness and transparency concerning national statistics. This included documents from NCSES, the Committee on National Statistics, the U.S. Office of Management and Budget (OMB), the United Nations Economic Commission for Europe (UNECE), Statistics Canada, the American Association for Public Opinion Research (AAPOR), and the White House.

The panel is also indebted to John Gawalt, previous director of NCSES, who not only helped to develop the funding for this study, but also served as unpaid consultant until May 2020. His knowledge of the federal statistical system and NCSES was invaluable as the panel interpreted its charge and organized its open sessions. In addition, John actively participated in weekly meetings or conference calls with the chair and staff which greatly helped clarify what issues the panel needed to focus its attention on and which helped organize the structure of the report.

The panel itself could draw on its own considerable expertise advising on programs from the federal statistical system, or in areas relevant to the new directions that had been discussed at a prior workshop on transparency. By subject area, these experts included: from federal statistical system: Philip Ashlock (General Services Administration, including data.gov), Christopher Chapman (National Center for Education Statistics), Dan Gillman (Bureau of Labor Statistics, Census Bureau), Dan Kasprzyk (Census Bureau, National Center for Education Statistics), Peter Miller (Census Bureau), and Sarah Nusser (Iowa State University); concerning metadata standards and tools: David Barraclough (Organisation for Economic Co-operation and Development [OECD]) and Dan Gillman; from international statistical agencies: David Barraclough (OECD), Frauke Kreuter (Joint Program of Survey Methodology and the University of

ACKNOWLEDGMENTS

Mannheim), and Eric Rancourt (Statistics Canada); concerning computer science tools applicable to federal statistics: H.V. Jagadish (University of Michigan), Audris Mockus (University of Tennessee), and Lars Vilhuber (Cornell University); concerning archiving: Margaret Levenstein (Inter-university Consortium on Political and Social Research) and Lars Vilhuber; and from the statistical user community: Linda Jacobsen (Population Reference Bureau).

In creating the chapters of our report, the following individuals played a key role: the first draft of the Summary was completed by Connie Citro of CNSTAT; Chapter 1 and the tables in Chapter 7 were primarily drafted by Peter Miller; Chapter 3 was primarily drafted by Lars Vilhuber, Margaret Levenstein, and Frauke Kreuter; important parts of Chapter 4 were drafted by Audris Mockus and Linda Jacobsen; Chapter 5 was drafted by Dan Gillman and David Barraclough, and sections of this chapter were drawn from material provided by Michael Lenard and Andrea Thomer, both of the University of Michigan, consultants to the panel. Under the panel's guidance, Lenard and Thomer also completed the first draft of Appendix A, while Dan Gillman drafted Appendix B.

Finally, the panel thanks staff for the preparation of the entire report. Michael Cohen and Michael Siri provided tireless energy and enthusiasm to the panel and its work, organizing open meetings, individual phone calls, and Zoom meetings, following up on a myriad of issues and comments, and organizing and drafting the report. Following through on the comments and ideas of panel members was a significant undertaking. The panel appreciated their interest and effort. Jillian Kaufman and Anthony Mann provided excellent administrative support during the panel's data gathering activities.

This Consensus Study Report was reviewed in draft form by individuals chosen for their diverse perspectives and technical expertise. The purpose of this independent review is to provide candid and critical comments that will assist the National Academies of Sciences, Engineering, and Medicine in making each published report as sound as possible and to ensure that it meets the institutional standards for quality, objectivity, evidence, and responsiveness to the study charge. The review comments and draft manuscript remain confidential to protect the integrity of the deliberative process.

We thank the following individuals for their review of this report: Katharine G. Abraham, Joint Program in Survey Methodology, University of Maryland, College Park; Christopher Carrino, Office of the Chief Information Officer, U.S. Census Bureau; Leighton L Christiansen, Bureau of Transportation Statistics; Mick P. Couper, Institute for Social Research, University of Michigan; Robert L. Griess, Department of Mathematics, University of Michigan; Pascal Heus, Metadata Technology North America; Nicholas Horton, Statistics and Data Science, Amherst College; Juan

Muñoz López, Informatics Planning and Governance, National Institute of Statistics and Geography of Mexico (INEGI); Regina L. Nuzzo, Freelance Science Writer, Washington, DC; and Nancy A. Potok, Chief Statistician of the United States (retired).

Although the reviewers listed above provided many constructive comments and suggestions, they were not asked to endorse the conclusions or recommendations of this report nor did they see the final draft before its release. The review of this report was overseen by Alicia L. Carriquiry, Department of Statistics, Iowa State University, and Roderick J.A. Little, Department of Biostatistics, University of Michigan. They were responsible for making certain that an independent examination of this report was carried out in accordance with the standards of the National Academies and that all review comments were carefully considered. Responsibility for the final content rests entirely with the authoring committee and the National Academies.

Daniel Kasprzyk (*Chair*)
NORC at the University of Chicago

Contents

Summary   **1**

1 Introduction   **15**
Definitions of Transparency and Reproducibility, 18
Practical Benefits of Transparency, 21
Calls for Transparency, 22
Some Constraints, 30
Report Structure, 31

2 Current Practices for Documentation and Archiving in the Federal Statistical System   **33**
The Complexity and Scientific Nature of the Production of Official Statistics, 33
Why Transparency and Reproducibility Are Goals for NCSES and the Federal Statistics System, 35
Existing Requirements, 38
Existing Practices, 39
Responses to the Informal Questionnaire, 40
Implications of Informal Questionnaire Results, 44
Challenges That Arise in Implementing Transparency and Reproducibility, 46

3 Changes in Archiving Practices to Improve Transparency   **51**
Transparency and Archives, 52
Archiving History and Practices, 54

Current Practices with Record Schedules and Data Management Plans, 58
The Role of Catalogs and Searchable Metadata, 62
Issues Arising with Paradata, 66

4   Assessments of Quality, Methods for Retaining and Reusing Code, and Facilitating Interaction with Users   73
Introduction, 73
Assessing the Quality of Inputs Used to Produce Official Estimates, 75
Transparency in Processing, Software Development, 82
Facilitating User Interaction with Statistical Agencies, 89

5   Metadata and Standards   95
Introduction, 95
Metadata: The Basics, 95
Metadata Systems, 104
Risks and Benefits, 107
Using Existing Systems, 111
Standards and Interoperability, 116
Examples of Statistical Metadata Standards, 122
Conclusion, 128

6   Making the Practices of the National Center for Science and Engineering Statistics More Transparent   131
Description of NCSES Programs, 131
Transparency for External Users of NCSES Survey Output, 138
Ease of Use of Information for Analysis Purposes, 141
Priorities for NCSES, 142

7   Best Practices for Federal Statistical Agencies   147
Best Practices for Documentation, Retention, Release, and Archiving of Data, 147
Dealing with Errata in Official Statistics, 165
A Vision of Federal Statistics in the Future, 166
Resource Needs to Proceed, 168

References   171

Appendixes

A Statistical Metadata Standards—in Detail   177
B The Role of Metadata in Assessing the Transparency of
   Official Statistics   219
C Public Meeting Agendas   227
D Biographical Sketches of Panel Members   233

Boxes, Figures, and Tables

BOXES

S-1 Benefits of Transparency to Federal Statistical Agencies, 4

1-1 Statement of Task, 18

2-1 Programs That Responded to Informal Panel Questionnaire, 40

3-1 Recent Classification Issue at the Bureau of Labor Statistics, 56
3-2 NCSES and Paradata, 61
3-3 Excerpts from 44 U.S. Code § 3511: Data inventory and Federal Data Catalogue, 64
3-4 Examples of Guidelines for the Retention of Paradata, 70

FIGURES

5-1 Example of a simple dataset description in XML, 101
5-2 A simple dataset description in RDF, 103
5-3 Conforming to standards—efficiencies gained, 121

A-1 GSBPM: Its processes, phases, and sub-activities, 182
A-2 BLS business process model, 184
A-3 GSIM top-level groups, 186
A-4 Simplified view of GSIM, 187
A-5 Alternate but simplified view of GSIM, 188

A-6   How GSIM and GSBPM work together, 189
A-7   GSBPM levels implemented in GSIM, 190
A-8   Overview of capabilities and (conceptual) building blocks of CSDA, 198
A-9   Data life cycle as conceived in DDI Data Lifecycle, 201

TABLES

1-1   OMB Standards and Guidelines for Statistical Surveys: Sections 7.3 and 7.4, 27
1-2   U.S. Census Bureau's Statistical Quality Standard F2: Providing Documentation to Support Transparency in Information Products, 29

6-1   NCSES' Survey Portfolio, 132

7-1   Documenting Basic Elements of a Statistical Program, 149
7-2   Documenting Statistical Programs Using Survey Data, 151
7-3   Documenting Statistical Programs Using Administrative Records and/or Digital Trace Data, 157
7-4   Documenting Data Integration Issues, 161
7-5   Documenting Paradata from Statistical Programs, 163
7-6   Archiving of Data, 164

A-1   CSDA Principles: Statements, Rationales, and Implications, 182

B-1   Elements, 223
B-2   Elements for Describing Variables, 224
B-3   Extended Elements for Describing Variables, 224

Acronyms and Definitions

AAPOR	American Association for Public Opinion Research
API	application programming interface
BEA	Bureau of Economic Analysis
BLS	Bureau of Labor Statistics
BTS	Bureau of Transportation Statistics
CAPI	computer-assisted personal interview
CATI	computer-assisted telephone interview
CE	Consumer Expenditure Survey
CNSTAT	Committee on National Statistics
CSDA	Common Statistical Data Architecture
CSPA	Common Statistical Production Architecture
DCAT	Data Catalog Vocabulary [DCAT] [related: DCAT-US, DCAT-AP]
DDI	Data Documentation Initiative
DMP	Data Management Plan
DSD	Data Structure Definition
ECDS	Early Career Doctorates Survey
EIA	Energy Information Administration
FAIR	Findable, Accessible, Interoperable, and Reusable
FCSM	Federal Committee on Statistical Methodology

FSRDC	federal statistical research data center
GPS	Global Positioning System
GSBPM	Generic Statistical Business Process Model
GSIM	Generic Statistical Information Model
HLG-MOS	High Level Group for the Modernization of Official Statistics
ICPSR	Inter-University Consortium for Political and Social Research
ICSP	Interagency Council on Statistical Policy
ISO	International Organization for Standardization
JSON	JavaScript Object Notation
LEHD	Longitudinal Employer-Household Dynamics
MEPS	Medical Expenditure Panel Survey
NARA	National Archives and Records Administration
NASS	National Agricultural Statistical Service
NCES	National Center for Education Statistics
NCHS	National Center for Health Statistics
NCSES	National Center for Science and Engineering Statistics
NSCG	National Survey of College Graduates
NSF	National Science Foundation
OECD	Organisation for Economic Co-operation and Development
OMB	U.S. Office of Management and Budget
PII	personally identifiable information
PUMD	public use microdata
RDAS	Restricted Data Analysis System
RDF	Resource Description Framework
SDMX	Statistical Data and Metadata eXchange
SDR	Survey of Doctorate Recipients
SIS-CC	Statistical Information System Collaboration Community
SSDC	Survey Sponsored Data Center
UML	Unified Modeling Language
UNECE	United Nations Economic Commission for Europe
URI	Uniform Resource Identifier

W3C   World Wide Web Consortium

XML   eXtensible Markup Language

Administrative records data: Data held by agencies and offices of the government that have been collected for other than statistical purposes to carry out basic administration of a program. (US OMB 2014 Guidance for Providing and Using Administrative Data for Statistical Purposes M-14-06.)

Archive: The National Space Science Data Center of the National Aeronautics and Space Administration (NASA) defines archives as follows (emphasis added):

> The term 'Archive' has come to be used to refer to a wide variety of storage and preservation functions and systems. Traditional Archives are understood as facilities or organizations which preserve records, originally generated by or for a government organization, institution, or corporation, for access by public or private communities. *The Archive accomplishes this task by taking ownership of the records, ensuring that they are understandable to the accessing community, and managing them so as to preserve their information content and Authenticity.* ...The major focus for preserving this information has been to ensure that they are on media with long term stability and that access to this media is carefully controlled. (p. 2-1)[1]

Data management plans: A data management plan is a knowledge management document, prepared initially as a specific research or survey project is being planned, to lay out types of data to be collected, the possible presence of sensitive data, the roles of project members in relation to the data, and the planned archiving and preservation of the data. A data management plan can be a living document that may change many times over the course of the research or survey project. (https://www.usgs.gov/products/data-and-tools/data-management/data-management-plans)

Digital trace data: This includes data collected via the Internet to represent transactions of various kinds, grocery store scanner data, data collected to record mobile phone activities, data from radio frequency identification tags, etc.

Discoverability: Discoverability is the use of standard metadata to describe one's datasets in a structured way, which makes it more likely that search

---

[1] Management Council of the Consultative Committee for Space Data Systems, 2012.

engines will be able to link these structured metadata with information describing its location and provide other linkages such as scientific publications and thereby facilitating its discovery for others.

Machine-actionable metadata: Machine-readable metadata in a format that can be used to drive some processes. This generally means there are no free-text fields. Fields that might be open text are instead populated by codes associated with a controlled vocabulary of possible entries.

Machine-readable metadata: Metadata in a format that can be read by a computer. The implication is that each metadata field may be individually separated and read. Documents rendered in HTML or PDF are readable by a computer program, but there are no individually readable fields.

Metadata: Data being used to describe some object(s). **Statistical metadata** are data (information) used to describe statistical objects, i.e., the metadata associated with a dataset, including the origins of the data, assessments of its quality, the variables included, their context and definitions, their values, their location in the database, what the different cases in the file refer to, and so on. Statistical metadata are best understood and most useful as structured information. Statistical metadata should be sufficient to allow someone not involved in an official statistics program to properly analyze an archived dataset resulting from that program. As Vardigan and Whiteman (2007) point out:

> for a secondary analyst to understand a given dataset, he or she must have access to good documentation ... A data file is ultimately just a string of numbers and not understandable on its own; it can only be interpreted and comprehended intellectually through use of the technical documentation ... which indicates a variable's location in the numeric data file, the question it was based on, all possible responses to the question, how the population of interest was sampled (for surveys) and so forth. (p. 76)

Metadata standard: A standard that addresses the kinds, meaning, and/or structure of data used as metadata. Standards are built through a consensus process that is open (any interested stakeholder may join), fair (every participating stakeholder has the same rights and privileges), observable (the process is open for inspection), and balanced (the participating stakeholders are representative of the entire set).

Metadata tool: A system developed for accessing or using metadata. Tools may be commercial, open source, or agency built. They are designed to address at least one aspect of the life cycle of metadata. Tools built to be used with a metadata standard are more widely applicable, since they can be adopted by any agency using that standard.

Paradata: "[A]dditional data that can be captured during the process of producing a statistic" (Kreuter, 2013). Such data are obtained throughout the survey process—as part of the initial interaction, the field staff's observations, and the respondent's actions. The data can be used to help ascertain and improve the quality of the collected data. Paradata, in the context of official statistics, are mainly used in conjunction with survey data and may consist of any information that helps to assess the ability of the respondent to respond accurately to the items in a (survey) instrument. What paradata will be collected for administrative records data or digital trace data is currently a research topic.

Record schedules: 36 CFR Subchapter B - RECORDS MANAGEMENT All Federal records, including those created or maintained for the Government by a contractor, must be covered by a NARA-approved agency disposition authority, SF 115, Request for Records Disposition Authority, or the NARA General Records Schedules. (36 CFR § 1225.10) General Records Schedules (GRS) are schedules issued by the Archivist of the United States (NARA) that authorize, after specified periods of time, the destruction of temporary records or the transfer to the National Archives of the United States of permanent records that are common to several or all agencies. (36 CFR § 1227.10) All agencies must follow the disposition instructions of the GRS, regardless of whether or not they have existing schedules.

Summary

Widely available, trustworthy government statistics are essential for policy makers and program administrators at all levels of government, for private-sector decision makers, for researchers, and for the media and the public. In the United States, 13 principal statistical agencies as well as units and programs in many other agencies produce various key statistics in areas ranging from the science and engineering enterprise to education and economic welfare. Their work is coordinated by the chief statistician in the U.S. Office of Management and Budget (OMB) and the Interagency Council on Statistical Policy (ICSP).

Official statistics are often the result of complex data collection, processing, and estimation methods. These methods can be challenging for agencies to document and for users to understand—they are not intrinsically transparent (a term defined below). The importance of openness of official statistics is noted in the Committee on National Statistics' *Principles and Practices for a Federal Statistical Agency* (NASEM, 2021):

> Federal statistical agencies must have credibility with those who use their data and information. . . . Because few data users have the resources to verify the accuracy of statistical information, users rely on an agency's reputation . . . . Agencies build and maintain respect and trust through clear public commitments to professional practice and transparency in all that they do, including informing users of the strengths and weaknesses of their data [emphasis added].

Principles and Practices is not the only high-level source calling for transparency in official statistics. Other sources include OMB's *Standards and Guidelines for Statistical Surveys,* the Federal Data Strategy, and the Foundations for Evidence-Based Policy Act. These important sources all recognize the importance of transparency and the closely related concept of reproducibility, and they all call for agencies to strive to achieve transparency in their work.

TRANSPARENCY AND REPRODUCIBILITY

The panel could find no formal definition of the term *transparency* when used in conjunction with official statistics, though its desirability is often cited. Two cases in which the transparency of official statistics is touched upon are the following. First, the Quality Assurance Framework of the European Statistical System Version 2.0 contains this language:

> Transparency of processes. The statistical authorities document their production processes and documentation of these processes is available to staff. A condensed/summary version is made available to users through user-oriented quality reports based on ESS standards, i.e. Single Integrated Metadata Structure (SIMS).

Second, the UN Statistics Quality Assurance Framework, which defines transparency in conjunction with objectivity, impartiality, and professionalism, states that transparency is ...

> publicising the methods used [and] ... ; ensuring that statistics are determined by statistical considerations and not by pressure from providers or users and explaining major changes in methodology to users.

This report has drawn on existing definitions of *reproducibility*, and defines what transparency and reproducibility mean in this particular context. For our purposes, transparency is the provision of sufficiently detailed documentation of all the processes of producing official estimates. The goal of transparency is to enable consumers of federal statistics to accurately understand and evaluate how estimates are generated. There are different levels of understanding. Since consumers vary in their interests and needs, transparent documentation includes basic information for the merely curious observer as well as technical information for experts. Similarly, there are different levels of evaluation, ranging from individual impressions about the usefulness of official estimates for idiosyncratic purposes to the most rigorous form of assessment, an attempt to reproduce the estimates in an independent investigation. Transparency makes it possible to understand how official estimates came to be as they are, and whether they are reliable.

While investigations of the reliability of federal statistics by outside organizations are not common, it is essential for agencies to have available the information necessary for them to be undertaken. The credibility of official estimates is undermined if questions about their genesis cannot be answered or if it is impossible to check them. The recent National Academies of Sciences, Engineering, and Medicine report on *Reproducibility and Replicability in Science* (2019b) observes that there are some general questions about the reliability of research results that have been raised across all scientific disciplines:

1. Are the data and analysis laid out with sufficient transparency and clarity that the results can be checked?
2. If checked, do the data and analysis offered in support of the result in fact support that result?
3. If the data and analysis are shown to support the original result, can the result reported be found again in the specific study context investigated?
4. Finally, can the result reported or the inference drawn be found again in a broader set of study contexts? (p. 44)

In this report, we are concerned with transparency in relation to the first three questions. Answering questions 1 and 2 involves scrutinizing and employing information from the statistical agency to check results that the agency has published. This sort of investigation—notably involving data analysis (including "cleaning," editing, and weighting) and associated computer code—seeks to determine if the published results based on these stages of the research process can be reproduced. Answering question 3 involves conducting a much broader independent investigation, within the "specific study context" that produced the original official estimates. The context is the full set of study components, from conceptualization to design to data collection to data analysis and publication.

Can conducting a parallel investigation, using the same procedures as those followed by the agency to construct the official estimates, reproduce those estimates, within a reasonable margin of error? Further, one should have a prespecified margin of error that one anticipates from reproducing estimates from independent studies. This idea is specified in reproducibility exercises done by people connected with the Center for Open Science. Recognizing that such studies would be very complex and expensive, our recommendations for transparent documentation are aimed at urging agencies to have available the information needed to make them possible.

Finally, transparency includes *discoverability*. Data users have to be able to readily locate sufficient information about current and past statistical programs and sets of estimates and input datasets to be both aware of their existence and to evaluate their quality and fitness for use.

Transparency in federal statistics has multiple benefits, as highlighted in Box S-1. First, transparency is consistent with sound management. If federal agencies understand that their internal processes are subject to external scrutiny, this will encourage thoroughness and care. Further, agencies can better cope with the inevitable temporary and permanent changes to staff if they retain, in an accessible way, detailed information as to how they accomplish their various data collection, data treatment, and estimation tasks in the production of official statistics.

Second, transparency facilitates innovation and improvement in statistical methods. Internal researchers from a statistical agency producing official statistics, as well as external researchers, are in a better position to enhance methods for data collection or estimation if clear and detailed documentation of existing processes and sources for producing official estimates is maintained and made accessible.

Third, transparency buttresses confidence in statistical methods. Federal statistics have broad impact: they are used to apportion political representation to states, to form congressional districts, and to allocate substantial governmental funds to regions. They are also used to inform and assess the effectiveness of policies to improve public health, education, the economy, employment, agriculture, and commerce. One critical component of trust—both of building trust and earning it—is for federal statistical agencies to

---

BOX S-1
Benefits of Transparency to Federal Statistical Agencies

1. Transparency is consistent with sound management. It is important for agencies to have a complete description of how their estimates are produced to support their review and to be able to accommodate staff changes.

2. Transparency supports innovation in and improvement of statistical methods.

3. Official statistics are used for important purposes, so it is important that they be trustworthy and known to be unbiased and of high quality. Along the same lines, transparency facilitates computational reproducibility.

4. Transparency increases data utility by supporting alternative tabulations and alternative analyses, including combining information approaches.

"open their books" to the extent feasible, that is, to be transparent as to the data they collect and the methods they use to produce the official statistics.

Along the same lines, transparency is closely aligned with the *reproducibility* of a set of official statistics, whose value for scientific activities has lately acquired renewed importance. Due to the difficulty of controlling precisely what happens during a survey interview, reproducibility—that is, the ability to repeat the entire process of planning and design, data collection, data treatment, and statistical estimation in support of the production of a set of official statistics—is difficult to attain. It therefore can be argued that one should assess reproducibility in this context in the same way it was defined in *Reproducibility and Replicability in Science* (NASEM, 2019b). The definition in that report focuses on reproducing the computational steps used to produce the findings or results. To be in agreement with that notion of reproducibility, statistical agencies would need to retain the code used for all data treatments and estimation steps. If one then used the archived input datasets as inputs into this retained code (in the appropriate computer environment) and arrived at the identical official statistics, this would support the statement that a given set of official statistics was (computationally) reproducible. However, we believe that this sense of reproducibility is not sufficient. The panel feels that any components of the planning and data collection processes that are repeatable, such as the use of the computer code developed in support of computer-assisted survey techniques, and the use of the survey instrument itself, also need to be documented and retained for review. Therefore, the panel argues for a sense of reproducibility that is more comprehensive than computational reproducibility.

Fourth, transparency increases data utility. Users of information collected by federal agencies have to be able to understand its strengths and limitations. Informed data use leads to sounder conclusions and policy formation. This is true whether data users are creating their own tabulations from a single data collection or combining information from several sources. As federal agencies augment their data collection efforts to include more use of administrative records and other nonsurvey data, transparent processes become even more important.

Recognizing the need to enhance the transparency and reproducibility of its statistics, for all of the above reasons, the National Center for Science and Engineering Statistics (NCSES) asked the National Academies' Committee on National Statistics to establish a consensus Panel on Transparency and Reproducibility of Federal Statistics to study issues of documentation and archiving of NCSES statistical data products.

In addition, NCSES asked the panel to consider how NCSES could work with other federal statistical agencies to facilitate the adoption of currently available documentation and archiving standards and tools. The use of metadata standards and tools has been widely found to assist in both the documentation of methods and the archiving of official statistics and the input data used in their creation. In addition, if U.S. federal statistical agencies and various international statistical offices are to benefit from sharing methods, data, and results with each other, such efforts would be eased by making use of common tools for documentation and exchange.

TRANSPARENCY AS A NECESSARY COMPONENT OF FEDERAL STATISTICS

The panel reviewed federal legislation, OMB statistical policy directives and memoranda, best practice documents, including *Principles and Practices*, and other sources to identify what is variously required and recommended for running an effective federal statistical agency. The panel concluded that an effective agency should embrace transparency by providing important information that demonstrated independence from political and other undue external influence, supporting its consistency with current state-of-the-art methods, and exhibiting respect for and protection of data providers.

In its investigations, the panel came to the conclusion that the agencies of the federal statistical system are reasonably complete in what they retain internally for their agency. This, however, is not as true regarding the archiving of input datasets. Externally, agencies are not as transparent with respect to what is available to the public regarding data treatments and the methodologies employed. Also, agencies often do not provide information to researchers about what access is permitted and available via secure environments regarding input datasets.

> **Conclusion 2.1:** Documentation of data collection methods, data treatments, and estimation methods by federal statistical agencies, while in need of some improvement, is generally fairly complete with respect to what is available internally to an agency. The practice of archiving input datasets and official estimates varies greatly across agencies, and as a result some data are not retained even internally for long periods of time. Externally, while the public sometimes can gain access even to the code for various methodological processes, agencies often do not provide accessible methodological summaries for nonspecialists. Further, access to input datasets using secure avenues varies substantially across agencies.

It is clear from a wide variety of sources that transparency is an important goal to strive for in federal statistics, and that it is needed regarding the operations used, the methods applied, and the results obtained. There are several benefits to being transparent, but the most important are supporting greater trust that the official estimates are produced in an unbiased manner and supporting greater trust that they are of high quality.

Conclusion 2.2: A foundational element of agencies that produce federal statistics is transparency of operations, methods, and results so that users can trust that federal statistical estimates are produced in an unbiased manner and understand their properties and how best to use them. The principle of transparency is reinforced in numerous reports, directives, and legislation, including the Foundations for Evidence-Based Policymaking Act of 2018 and the Federal Data Strategy.

Given the importance of greater transparency of federal statistics, we offer the following recommendation.

Recommendation 2.1: Leadership at the Office of Management and Budget, the Interagency Council on Statistical Policy, the National Center for Science and Engineering Statistics, and all agencies that produce federal statistics should establish transparency of processes and methods as a high priority and continuously reinforce this priority to their staffs.

ARCHIVING PRACTICES

Federal law requires agencies to maintain record schedules, to preserve records considered to have long-term value in the National Archives of the United States, and to destroy records lacking value after a specified period.

Recommendation 3.1: The agencies that produce federal statistics, through the leadership of the Interagency Council on Statistical Policy and the Chief Statistician of the United States, should fully comply with federal record schedules, ensuring that the input datasets that can legally be retained, and official estimates that are produced, are archived in the National Archives and Records Administration. The metadata that accompany such data should also be preserved using broadly accepted metadata standards appropriate to the data at hand. The records schedules, which describe the plans for retaining, preserving, and making accessible microdata and associated metadata, should be easily accessible on each statistical agency Website so that users know when and where microdata and associated metadata will be made available, and when they are scheduled to be destroyed.

By doing this, the federal statistical agencies can then leverage their records schedules to improve transparency in their statistical programs.

Recommendation 3.2: Federal statistical programs, whose inputs include survey data, should make available, for as long as the data are believed to be of interest to researchers, associated paradata to help users assess the quality of the survey inputs.

When such paradata are associated with a statistical program that is used to distribute political power or substantial federal funds (such as the Decennial Census), and the paradata are a key measure of the quality of inputs to such a program, statistical agencies should make public such assessments for relatively disaggregated demographic-geographic domains.

TOOLS TO FACILITATE DOCUMENTATION OF METHODOLOGICAL PROCESSES

There are several tools widely used in academia and in industry that can facilitate the development and documentation of software processes. Since the federal statistical agencies use software processes to help to collect data, treat data in preparation for use in estimation, and estimate a set of official statistics, these tools should be examined for their utility in supporting greater transparency of official statistics.

Recommendation 4.1: Agencies that produce federal statistics, including the National Center for Science and Engineering Statistics, should review and make a priority of adopting modern information technology tools that assist in collaborative software development and documentation of workflow and methodology.

This last recommendation is important, because transparency of computational processes is as important as the transparency of other processes, and also because it will make their transparency efforts more efficient.

Recommendation 4.2: To facilitate transparency, agencies that produce federal statistics are encouraged to develop coding style guides, and to make available documentation and specifications for software systems, subject to any security concerns. Where possible, code (for example used for data collection or processing) should be made publicly available, subject to redaction or removal of confidential parameters, and logs of processing sequences should be archived. Manual processing steps should be clearly identified and documented, and any instructions or guidance given to the staff conducting such manual processing should be archived and made as transparently available as possible.

ISSUES CONCERNING TRANSPARENCY AND METADATA STANDARDS

The panel identified several aspects of the use and development of metadata standards relevant to federal statistical agencies. In addition, the federal statistical system as a whole, including NCSES, needs to lead in framing systemwide standards for the documentation and archiving of official statistics. This work should review and incorporate, where appropriate, international standards and processes that are widely in use, and it should provide training for relevant employees.

> **Recommendation 5.1:** The Interagency Council on Statistical Policy should develop and implement a multi-agency pilot project to explore and evaluate employing existing metadata standards and tools to accomplish data sharing, data access, and data reuse. The National Center for Science and Engineering Statistics should be an active agency participant in the project.

Such a pilot project could start with programs within agencies and agencies entering into more data sharing and data reuse agreements through the use of current methods, and then proceeding incrementally, assessing how sharing is facilitated through the use of existing metadata standards and tools. When one agency makes some of its data discoverable, this contributes to another agency's success in being transparent about its own inputs. Therefore, this greater emphasis on data sharing and data reuse is another motivation to make data discoverable.

In addition to such pilot projects, the leadership of the federal statistical system should encourage its agencies to play a more active role in the development of metadata standards and tools.

> **Recommendation 5.2:** The Interagency Council on Statistical Policy should (1) prioritize and emphasize the importance and benefits of federal statistical agency staff engaging in international metadata standards and tool development, and (2) organize a discussion among statistical agencies that leads to an effective, coordinated, and accountable approach for staff in agencies that produce federal statistics to contribute to international metadata standards and tool development.

Finally, the Chief Statistician and the ICSP should develop a continuing education program for agency staff about the nature and purpose of federal government requirements:

Recommendation 5.3: The Interagency Council on Statistical Policy and the Chief Statistician of the United States should develop a continuing education program for agency staff on the nature and purpose of federal government requirements for statistical activities (such as the Foundations for Evidence-Based Policymaking Act and the Federal Data Strategy) that have been issued, the expected positive return to the agency for their implementation, and the need for transparency in agencies that produce federal statistics.

RECOMMENDATIONS FOR NCSES

As part of its investigations, the panel reviewed NCSES's general publication standards as well as information available on NCSES's Web pages for four programs. These four were chosen because they included both individuals and institutions as respondents, at least one program involved data collected by the Census Bureau, and another involved data collected by a private contractor.[1] In all four cases, the survey design was well described. However, more than one program requested users to go to a technical report to find out what data treatments were used to address nonresponse or failed edits. Further, it was not always clear where the input data or resulting official estimates were archived.

The panel also heard from two expert users of NCSES data, who praised the value of NCSES data but indicated ways in which they could be made more useful for various types of analysis. The panel also learned that NCSES has recently created considerably improved user interfaces for users to prepare customized tables. Based on this, the panel identified several priority areas for NCSES to address so that its data series can be more transparent and helpful to users.

Recommendation 6.1: Agencies that produce federal statistics should word their contracts and interagency agreements that involve planning, collection, processing, and analysis of data products so that any information adhering to the agency's transparency expectations (e.g., all processes and code related to data handling, and the resulting data collected) that is obtained by the contractor or federal agency should be provided to the sponsoring agency unless constrained by legal or proprietary considerations.

---

[1] The four programs were Business Enterprise Research and Development Survey, Survey of Earned Doctorates, Higher Education Research and Development Survey, and Early Career Doctorates Survey.

Recommendation 6.2: The National Center for Science and Engineering Statistics' (NCSES) information technology staff and NCSES's program staff should collaborate to develop an *ongoing program* to seek user input to improve the functionality of their Web interface, to test new analytic tools, to make it easier for users to identify documentation resources, and to facilitate appropriate access to data.

Recommendation 6.3: The National Center for Science and Engineering Statistics' (NCSES) senior management should renew their emphasis on the timely production, distribution, and accessibility of methodology reports, data quality assessments, and quality profiles for each program, to ensure transparency about data quality and the information underlying NCSES's statistical programs. These studies on methodology and data quality should be a regular component of NCSES's ongoing work and be made available and accessible through the Website associated with each program.

Recommendation 6.4: The National Center for Science and Engineering Statistics' (NCSES) senior management should monitor and reinforce their agency's policy for archiving microdata that are the basis for the production of official statistics, as well as official statistics themselves. These policies should specify the use of data management plans, above and beyond legally required records schedules, which explicitly describe how the microdata and official statistics will meet FAIR (Findability, Accessibility, Interoperability, and Reusability) principles. The records schedule for a statistical program should always be easily accessible. The retained data and statistics should be available (via links, search capability) from the program's Website, to the extent possible consistent with confidentiality protections.

Recommendation 6.5: The National Center for Science and Engineering Statistics' (NCSES) information technology staff and NCSES's program staff should collaborate to standardize the inclusion of language in their contracts and interagency agreements requiring that contractors provide machine-actionable metadata and code so that NCSES can meet acceptable standards of transparency about its data products for users and other agencies and achieve consistency in the metadata used across NCSES's statistical programs. The NCSES chief statistician should monitor the implementation of this policy.

Recommendation 6.6: The National Center for Science and Engineering Statistics' (NCSES) senior management should investigate how their programs use paradata (process data, such as the number of interview contact attempts, the interim and final case dispositions, the duration of completed interviews), identify programs that would benefit from the use of paradata, identify what paradata are valuable to maintain, determine the length of time such data can be made available to researchers, and ensure that records schedules include the status of such data. While individual programs have different requirements and uses, for the purpose of transparency NCSES management should develop a policy concerning the availability and use of paradata consistent with its mission.

Recommendation 6.7: Given the varying needs and expertise of different users, transparency is enhanced when all National Center for Science and Engineering Statistics' (NCSES) data programs take steps to help users interact with the data used to develop official statistics. NCSES should
- Establish ongoing data user groups with contact mechanisms;
- Establish a repeated survey of users as to their current experiences in accessing and using agency data and how estimates could be presented to facilitate time-series and cross-sectional analyses;
- Ensure consultations with data users prior to making changes in dissemination systems, statistical programs, and time series;
- Create a mechanism that enables members of a statistical program's user group to communicate directly with one another;
- Organize regular meetings with a broad user community representation; and
- Through surveys and direct interactions with users, identify ways to improve the transparency, accessibility, and usability of NCSES estimates, data products, documentation, and dissemination systems, including the structure and navigation of the agency's Website.

SYSTEMWIDE BEST PRACTICES

The panel tried to assess the degree to which the information currently provided by NCSES and the other federal statistical agencies about their programs and series of official statistics is consistent with the goals of being transparent. Many agencies do not have formal guidelines or rules that their statistical programs can use to decide what information to provide both internally to agency staff and publicly to their user communities. Instead, these decisions are often made at the programmatic level. Consequently, the panel sent an informal questionnaire to 11 high-profile federal statistical programs, asking appropriate staff to describe their program documentation and archiving practices. The panel also examined the Web pages associated with these programs.

The panel concluded that in general, agencies make a serious effort to provide information to enable users to understand how estimates were produced, and they include substantial information about the quality of any survey data inputs. However, agencies often do not provide details about the methods used to produce the treated data, and they often limit their published assessments of the quality of the resulting estimates to discussions of sampling error (for survey inputs). Moreover, agencies often do not make it entirely transparent as to where the archived data inputs and outputs are stored and what access to them is possible.

Because documenting and archiving federal statistical series is complicated, the panel identified best practices that agencies can use to determine the extent to which their policies are consistent with transparency. These best practices are divided into tables, in Chapter 7, that provide direction for agencies on what information they should retain internally and what should be made available to the public by establishing documentation requirements for the basic elements of a statistical program (Table 7-1), programs that utilize survey data (Table 7-2), programs that utilize administrative and digital trace data (Table 7-3), issues that arise in data integration (Table 7-4), paradata (Table 7-5), and archiving of data used in the production of official statistics (Table 7-6).

The processes giving rise to the production of official statistics typically involve (1) data collection, often through a survey but sometimes through use of administrative records or digital trace data; (2) data treatments, often to address failed edits or nonresponse; and (3) estimation, the use of methodological procedures to produce the official estimates. Once estimates are produced, (4) their quality needs to be assessed. Based on that division into component processes, the panel provides the following recommendation on what agencies are to retain.

Recommendation 7.1: The National Center for Science and Engineering Statistics (NCSES) and all agencies that produce federal statistics should, to the fullest extent feasible, document their data collection methods, their data treatments, their estimation methodologies, and assessments of the quality of their official estimates, and they should archive their input datasets and their official estimates to support reproducibility and later reuse, as specified in the tables developed by the panel. To the extent possible, they should make as much of this information as possible available to their external user communities; for data treatments and estimation methodologies, they may do so through methodological overviews. They should provide reasons, such as legal or contractual constraints, for omitting items in the tables.

The goal of the agencies is to strive for reproducibility, which in addition to computational reproducibility includes the retention of the documentable portions of the processes used prior to and as part of data collection. So, for example, if a program collected data using computer-assisted survey methods, the computer code used in such techniques would also need to be retained. That is, a record of all processes, even the human components, would be retained since doing so is a basic step in understanding and measuring the reliability of a process, as one might find in quality standards such as the ISO 9000 family[2] or the Capability Maturity Model Integration.[3] Human components could include such things as field protocols, or the treatment of data collected during natural disasters. To this end, the notion of computational reproducibility has to be expanded to include pre-data collection and data collection processes.

This notion of reproducibility is a more comprehensive notion than transparency, since it involves the actual detailed code used to produce a set of official statistics, rather than, say, only a detailed report *about* the methodology applied.

> **Recommendation 7.2:** Senior management at the agencies that produce federal statistics should provide resources and staff support to help transform their current processes to incorporate the use of data sharing and reuse through use of metadata tools and standards. This entails support for pilot projects, additional training of existing staff, enlisting of assistance from experts through support contracts, and reconfiguring of existing processes.

> **Recommendation 7.3:** Agencies that produce federal statistics, in order to implement many of the recommended initiatives in this report, should be provided with additional funds to acquire the necessary training and information technology assistance, as well as cover any increased operational costs, to modify current processes to improve documentation and archiving in support of the greater transparency of official statistics.

---

[2] For more information see https://www.iso.org/iso-9001-quality-management.html.
[3] https://cmmiinstitute.com/.

1

Introduction

The principal U.S. statistical agencies[1] are tasked with informing the public on various aspects of the state of the country and its residents, including such characteristics as the population count of states, counties, and cities, energy consumption, farm production, the state of the economy, educational attainment, and employment. Often, these official estimates are used to support decisions about how governmental policies should be implemented or modified to improve various dimensions of the nation's welfare. Further, official statistics are used to allocate political power and to distribute federal funds. To provide information on these important matters, the statistical agencies produce official statistics on the status of these various aspects of the United States.

These estimates must be seen as trustworthy. One step toward achieving trust is to maintain an "open book" policy. While all the federal statistical agencies already agree with having such a policy, in principle, it

---

[1] There are 13 principal statistical agencies: the Bureau of the Census and the Bureau of Economic Analysis in the Department of Commerce; the Bureau of Justice Statistics in the Department of Justice; the Bureau of Labor Statistics in the Department of Labor; the Bureau of Transportation Statistics in the Department of Transportation; the Economic Research Service and the National Agricultural Statistical Service in the Department of Agriculture; the Energy Information Administration in the Department of Energy; the National Center for Education Statistics in the Department of Education; the National Center for Science and Engineering Statistics in the National Science Foundation; the National Center for Health Statistics and the Office of Research, Evaluation, and Statistics of the Social Security Administration in the Department of Health and Human Services; and the Statistics of Income Division of the Internal Revenue Service in the Department of the Treasury. In addition to these 13 agencies, a number of other agencies have smaller units that have major statistical responsibilities.

is often unclear to them how it should be put into operation, in particular how much detail should be provided. Nevertheless, it seems clear that, at the least, sufficient information must be made available so that it is clear that official statistics are meeting high standards of scientific quality and integrity.

Two concepts that have been applied to scientific arenas where openness is warranted are those of *transparency* and *reproducibility*. In this context, transparency and reproducibility are to be achieved through release of documentation on the plans, processes, datasets, computations, and estimation methodologies used, both through release of the official estimates and through release of the evaluation of the input data and the official estimates. The increased accountability that would result from such openness can assure the public that the data were collected without bias and that the methods used are consistent with the current state of the art of statistical science. In a sense, there is an implied contract. The data are collected with funding from taxpayers, and statistical agencies ask for cooperation in their collection of that data, and in response the agencies return high-quality, objective official estimates that enable the public to make informed decisions. Transparency is necessary for the public to determine whether the statistical agencies are keeping their side of the bargain.

As indicated, these questions about transparency and reproducibility in official statistics are part of a much broader set of questions about transparency and reproducibility in science. These questions resulted in the National Academies of Sciences, Engineering, and Medicine undertaking a series of forums on open science, whose goal was to elicit from participants a variety of ideas on ways to support greater openness in a wide range of scientific enterprises. The last activity in that series was a workshop with a narrower focus: openness concerning the generation and publication of official statistics. As a result of the discussions at that workshop (see *Methods to Foster Transparency and Reproducibility of Federal Statistics,* NASEM, 2019a), one attendee, John Gawalt, then director of the National Center for Science and Engineering Statistics (NCSES), thought it would be valuable to have a study examining the extent to which NCSES practices, and those of federal statistical agencies more broadly, are currently transparent, what benefits greater transparency might offer, what tools might facilitate greater degrees of transparency, what are the legal, administrative, and resource-based constraints on being more transparent, and, if considered desirable, what are the appropriate steps to increase the degree of transparency both in the near term and over time. His successor, Emilda Rivers, was instrumental in bringing this to fruition.

This interest in transparency and reproducibility in official statistics is not new—it is fundamental to the scientific method. While there is renewed interest in reproducibility as a result of recent research in particular scientific disciplines arguing that a large fraction of research is not reproducible, the need to provide support for scientific activities by being open about the methods used and the data that were collected is longstanding. We encourage the professional staff of the federal statistical agencies to view their work as involving the application of scientific methods to produce official statistics and that therefore, consistent with scientific work, the actions taken throughout the process of developing these statistics need to be made transparent. If staff do not have this view of their work, steps taken to instill this attitude would be helpful. In addition, there are other reasons to support greater transparency. As with any complicated manufacturing process, for the production of statistics it is important to have a complete workflow history documenting how data are collected, how they are treated, how estimations are carried out, and how the quality of official estimates is assessed. These complete workflow histories allow federal statistical agencies to better manage and innovate in the production of official statistics.

In response to the NCSES request, the National Academies established the Panel on Transparency and Reproducibility of Federal Statistics for the National Center for Science and Engineering Statistics; the complete statement of task is in Box 1-1.

As is clear from the statement of task, the panel undertook its work with a dual focus: the degree of transparency of practices at NCSES and the degree of transparency in all of the principal agencies of the federal statistical system. The two goals dovetail to a considerable degree, because NCSES's policies and processes are typical of those of the other agencies. Further, as a relatively small statistical agency, NCSES is positioned to be nimble and innovative and to share what it learns with the broader statistical community. In addition, one of the benefits of greater documentation of methods and archiving of data is that it allows methods and data to be shared and reused either within an agency or across agencies. So, while as an individual agency NCSES can benefit from sharing and reusing data and methods internally, such benefits are likely to be even broader when considered across the entire federal statistical system and internationally. This is particularly relevant to NCSES, since it regularly interacts with international statistical agencies and the Census Bureau (as a major data collection agency for NCSES surveys). Examining the entire federal statistical system for its policies regarding transparency of data and methods, and assessing how using the same tools and standards in packaging the data and methods can facilitate sharing them across agencies, should yield important benefits.

> **BOX 1-1**
> **Statement of Task**
>
> An ad hoc panel will study issues of documentation and archiving of statistical data products for the National Science Foundation's National Center for Science and Engineering Statistics (NCSES). The desired objective is to enable NCSES to enhance the transparency and reproducibility of the agency's statistics and facilitate improvement of the statistical program workflow processes of the agency and its contractors. The panel will consider such issues as
>
> 1. What documentation and archiving guidance, standards, and tools currently exist to assist NCSES to facilitate transparency and reproducibility? Which ones are most useful and feasible to implement? For censuses and sample surveys? For administrative records? For statistics that may be based on combinations of these and other data sources?
> 2. In what ways can the costs of transparency for NCSES and its data users be minimized and the benefits maximized?
> 3. How can NCSES obtain value from saving and using the history of statistical program workflow processes to facilitate validation and to guide improvement over time in various processes, such as editing and imputation?
> 4. What are best practices to foster transparency internal to NCSES through more comprehensive documentation and archiving of methods and data?
> 5. What are best practices to foster transparency external to NCSES while maintaining an appropriate degree of disclosure control for confidential microdata?
> 6. What requirements for documentation and archiving standards and tools should NCSES include in contracts with data collectors, such as the U.S. Census Bureau and private survey firms?
> 7. What are feasible implementation steps toward better documentation and archiving for NCSES in the next 2-3 years? What should be the goals for a longer-term R&D effort in this area?
> 8. How can NCSES work with other federal statistical agencies to facilitate adoption of documentation and archiving standards and tools in common?

DEFINITIONS OF TRANSPARENCY AND REPRODUCIBILITY

The panel could find no formal definition of the term *transparency* when used in conjunction with official statistics, though its desirability is often cited. Two cases in which the transparency of official statistics is touched upon are the following: (1) the Quality Assurance Framework of the European Statistical System Version 2.0, which contains this language:

> Transparency of processes. The statistical authorities document their production processes and documentation of these processes is available to staff. A condensed/summary version is made available to users through

user-oriented quality reports based on ESS standards, i.e. Single Integrated Metadata Structure (SIMS).[2]

and (2) the UN Statistics Quality Assurance Framework, which defines transparency in conjunction with objectivity, impartiality, and professionalism as follows:

> publicising the methods used [and] ... ; ensuring that statistics are determined by statistical considerations and not by pressure from providers or users and explaining major changes in methodology to users.[3]

For our purposes, transparency is the provision of sufficiently detailed documentation of all the processes of producing official estimates. The goal of transparency is to enable consumers of federal statistics to accurately understand and evaluate how estimates are generated. There are different levels of understanding. Since consumers vary in their interests and needs, transparent documentation includes basic information for the merely curious observer as well as technical information for experts. Similarly, there are different levels of evaluation, ranging from individual impressions about the usefulness of official estimates for idiosyncratic purposes to the most rigorous form of assessment, which is the attempt to reproduce the estimates in an independent investigation. Transparency makes it possible to understand how official estimates came to be as they are, and whether they are reliable.

While investigations of the reliability of federal statistics by outside organizations are not common, it is essential for agencies to have available the information necessary for them to be undertaken. The credibility of official estimates is undermined if questions about their genesis cannot be answered or if it is impossible to check them. The recent National Academies report, *Reproducibility and Replicability in Science* (2019b), observes that there are some general questions about the reliability of research results that have been raised across all scientific disciplines:

1. Are the data and analysis laid out with sufficient transparency and clarity that the results can be checked?
2. If checked, do the data and analysis offered in support of the result in fact support that result?
3. If the data and analysis are shown to support the original result, can the result reported be found again in the specific study context investigated?
4. Finally, can the result reported or the inference drawn be found again in a broader set of study contexts? (p. 44).

---

[2] European Statistical System (2019, p. 28).
[3] United Nations Statistics Division (2016, p. 39).

In this report, we are concerned with transparency in relation to the first three questions. Answering questions 1 and 2 involves scrutinizing and employing information from the statistical agency to check results that the agency has published. This sort of investigation—notably involving data analysis (including "cleaning," editing, and weighting) and associated computer code—seeks to determine if the published results based on these stages of the research process can be reproduced. Answering question 3 involves conducting a much broader independent investigation, within the specific study context that produced the original official estimates. By "context" here we mean the full set of study components, from conceptualization to design to data collection to data analysis and publication. Can conducting a parallel investigation, using the same procedures as those followed by the agency to construct the official estimates, reproduce those estimates, within a reasonable margin of error?

Further, one should have a prespecified margin of error that one anticipates from reproducing estimates from independent studies. This idea is specified in reproducibility exercises done by people connected with the Center for Open Science.[4] Recognizing that such studies would be very complex and expensive, our recommendations for transparent documentation are aimed at urging agencies to have available the information needed to make them possible.

In a study of the reproducibility of a set of official statistics that was derived from a sample survey, one would use the same definition of the target population (e.g., "adults 18-75 in the United States, living in noninstitutional quarters [not prisons, nursing homes, etc.]"), the same sample design (e.g., multistage area probability design using primary sampling units [metropolitan areas], census tracts within those, census blocks within those, blocks within those, clusters of dwelling units within those, and individuals sampled randomly within those). The idea is to use the identical sample design but with a new sample. One would have to keep to the same time period, but there are some details that are unclear, such as whether one would use the same respondents, the same interviewers for each respondent, and whether efforts would be made so that the indicator of nonresponse would be the same across replications. Then reproducibility is assessed against an acceptable margin of error, which is primarily sampling error. But as suggested in the argument above, it might also include other errors, depending on how closely the reproducing study's procedures can mimic those of the official study.

---

[4]For example, see *Evaluating the replicability of social science experiments in Nature and Science between 2010 and 2015*, published by the Center for Open Science at https://osf.io/pfdyw/.

PRACTICAL BENEFITS OF TRANSPARENCY

We believe that agencies engaged in the production of official statistics have the obligation to strive to be as transparent as possible due to the fact that official statistics are the product of a scientific activity which, by its nature, implies a responsibility to indicate what work was done. Further, federal statistics are used for important purposes and therefore there is a need to be clear with users about how they were produced. In addition to supporting this norm of openness, a variety of benefits are gained when the agencies that produce official statistics provide greater transparency. These benefits can be combined into four categories: (1) efficiency, (2) innovation and progress, (3) trust and confidence, and (4) value from the use of the data products.

Efficiency arises at an agency producing statistics when what is done to produce them is known so completely so that any temporary or permanent changes to staff due to resignations, retirement, or sickness can be easily accommodated. As part of any organization that undertakes or oversees complex processes, it is necessary for statistical agencies to retain, in an accessible way, detailed information as to how they accomplish their various data collection designs, data treatments, and estimation tasks as components of the production of their official statistics. Transparency therefore can be seen as being consistent with sound management. By retaining this information, an agency can ensure that new hires or transfers can quickly get up to speed regarding what is needed and where in the process it fits in. Further, this knowledge facilitates the identification of sources of problems should they become known. Finally, if staff understand that their work is subject to review, it encourages a thoroughness and care that is beneficial.

Innovation and progress result from internal staff and external researchers understanding in some detail how official statistics are produced and thereby being able to discover areas in need of improvement. So transparency supports methodological innovation. Both internal and external researchers are in a better position to enhance methods for data collection or estimation if clear and detailed descriptions of the processes used for producing official statistics exist. These benefits arise, in particular, when researchers external to an agency are given sufficient information to conduct research on improving the methods, since they understand what is done and they can also assess the fitness of the data or estimates for other applications.

Trust and confidence derive from a full understanding of how a set of estimates are produced and knowing that what is done is consistent with state-of-the-art procedures and is not used to benefit any particular stakeholder. Official statistics matter. They are used to allocate substantial governmental funds and they are used to inform and assess the effectiveness of policies to improve public health, education, the economy, employment, agriculture, and commerce. Given their broad impact, official statistics should be produced using the best possible science, with complete objectivity, under the constraints of minimizing respondent burden, and expending needed funds in a cost-efficient manner. Most importantly, the public must have confidence that all of this is the case.

One component—of both building trust and earning it—is for federal statistical agencies to "open their books" to the extent feasible. Having an accurate and complete description of how an official statistical series is developed gives external users, especially those with some relevant expertise, confidence that the agency is approaching the data collection and estimation problem with care and objectivity. Along the same lines, knowing which input datasets were used to produce a set of estimates, with the statistical methods used to prepare them for the estimation methodology as well as the details of the estimation methodology, promotes trust. In addition, an "open books" approach provides the raw material for a test of computational reproducibility, which can greatly support a sense of trust in a set of official statistics.

Finally, for the full *value* of statistics to be realized through their use, the quality of the statistical processes must be documented. If the quality of a set of official estimates is not documented, users are more likely to misuse the estimates by not understanding their limitations or by using them in combination inappropriately.

CALLS FOR TRANSPARENCY

The need for transparency by statistical agencies is argued for by Rancourt (2019, p. 549) in an insightful perspective:

> With society becoming more complex and increasingly digitized, transparency is more needed, and in fact requested more than ever by citizens. Transparency is a key enabling piece of trust and accountability. It is beneficial to both social acceptability and scientific integrity and is an integral part of quality processes. Transparency is the quality of an object or process through which one can see. The definition can be quite fluid and mean a number of things depending on the context, but in the world of Official Statistics, it means that approaches, methods, decisions, and information are made available to users, researchers, stakeholders and citizens.

Such calls for transparency are not new. A 1978 report from the U.S. Department of Commerce states:

> To help guard against misunderstanding and misuse of data, full information should be available to users about sources, definitions, and methods

used in collecting and compiling statistics, and their limitations. (U.S. Department of Commerce, 1978, p. 11)

Transparency is mandated by statistical agencies outside the United States as well. In their *Policy on Informing Users of Data Quality and Methodology,* Statistics Canada outlines its following policy on transparency in data and methods:

> Statistics Canada, as a professional agency in charge of producing official statistics, has the responsibility to inform users of the concepts and methodology used in collecting, processing and analyzing its data, of the accuracy of these data, and of any other features that affect their quality or 'fitness for use'....[5]

In the European Union, Eurostat states that its Code of Practice

> is the cornerstone of the common quality framework of the European Statistical System ... based on 16 Principles covering the institutional environment, statistical processes and statistical outputs...The development, production and dissemination of our statistics are based on sound methodologies, the best international standards and appropriate procedures that are well documented in a transparent manner.[6]

More recently, in the United States, the *Report of the Commission on Evidence-Based Policymaking* (2017)[7]—which the federal statistical agencies are using to guide them in updating their own programs and methods over the next several years—includes the following statements:

- Government also can dramatically improve transparency about its collection and use of data, improving the American public's ability to hold the government accountable. Adhering to the highest possible standards with respect to privacy and accountability is an important part of earning the public's trust. (p. 8)
- Transparency. Those engaged in generating and using data and evidence should operate transparently, providing meaningful channels for public input and comment and ensuring that evidence produced is made publicly available. (p. 17)
- The existing infrastructure for accessing, linking, and analyzing confidential data for evidence building does not always prioritize state-of-the-art transparency and oversight. (p. 74)

---

[5] https://www.statcan.gc.ca/eng/about/policy/info-user.
[6] https://ec.europa.eu/eurostat/web/quality/european-statistics-code-of-practice.
[7] https://bipartisanpolicy.org/commission-evidence-based-policymaking/.

Growing out of the work of the Commission, the Foundations for Evidence-Based Policymaking Act of 2018 (P.L. 115-425, a subset of which is the OPEN Government Data Act) indicates how agencies should implement many of the recommendations of the Commission. The Act contains two sections relevant to the panel's work:

> (101) The bill requires agencies to submit annually to the Office of Management and Budget (OMB) and Congress a systematic plan for identifying and addressing policy questions. The plan must include, among other things ... data the agency intends to collect, use, or acquire to facilitate the use of evidence in policymaking; methods and analytical approaches that may be used to develop evidence to support policymaking; ... Agency strategic plans shall contain an assessment of the coverage, quality, methods, effectiveness, and independence of the agency's statistics, evaluation, research, and analysis efforts.
>
> ...
>
> (202) This bill requires public government data assets to be published as machine-readable data. The General Services Administration must maintain an online federal data catalogue to provide a single point of entry for the public to access agency data. Each agency shall develop and maintain a comprehensive data inventory.

In addition, in March 2018, "The President's Management Agenda" laid out a new cross-agency priority goal: leveraging data as a strategic asset to develop and implement a comprehensive Federal Data Strategy. One of the principles of this strategy is to promote transparency. The Federal Data Strategy also urges the adoption of 40 practices, of which eight (strategies 5, 14, 16, 19, 20, 26, 29, and 33) are relevant to this report.[8]

The Panel wishes readers to see that the views expressed in the reports of these two advisory groups motivate many of the recommendations that are offered here and that implementing these recommendations will result in a more modern federal statistical system, one that uses efficient processes and whose methods and data are more sharable across agencies.

Finally, an early 2021 memorandum (January 27th) from the Biden administration, entitled "Restoring Trust in Government Through Scientific Integrity and Evidence-Based Policymaking,"[9] addressed, among other things, the need for transparency and reproducibility of federal statistics. This memorandum follows other recent official memos emphasizing the need for better record retention (at the National Archives and Records Administration), especially administrative records data, in order to support greater reuse of collected data both by researchers and by other statistical agencies.

---

[8] https://strategy.data.gov/action-plan/.

[9] https://www.whitehouse.gov/briefing-room/presidential-actions/2021/01/27/memorandum-on-restoring-trust-in-government-through-scientific-integrity-and-evidence-based-policymaking/.

INTRODUCTION

There are also recent instances when some failure to document methods or archive input datasets has resulted in a loss of important information. One example (among many) reported on during the 2017 workshop on transparency that preceded this study (NASEM, 2019a) is that the Census Bureau did not retain documentation of the entire workflow used in conducting the 2010 Census. Although the Census Bureau did retain many component processes and data, failure to retain the entire workflow made it more difficult to evaluate some of those processes, which may have affected preliminary work in designing the 2020 census.

Although we did not undertake a formal investigation, we suspect that in the federal statistical system documentation of methods and retention of input datasets is less complete for one-time programs than for continuing programs. In addition, there are concerns that retention of methods and data is often carried out in a way that does not facilitate sharing such techniques and data among agencies, due to the fact that the formats used for the descriptive metadata are often agency specific. Further, much of what is done is survey-centric, so it is much less clear what information ought to be retained, even internally, for programs making use of administrative records or digital trace data. (Digital trace data are data that are collected in concert with our use of various forms of technology, especially the Internet but also, e.g., data on smartphone use and supermarket scanner data.)

Relevant Transparency Initiatives at the Office of Management and Budget, the Census Bureau, and the American Association for Public Opinion Research (AAPOR)

OMB's Standards and Guidelines for Statistical Surveys.[10] Produced by OMB in 2006, *Standards and Guidelines for Statistical Surveys* provides extensive instruction for statistical agencies. Of the standards it prescribes, three relevant to this report are reproduced below:

> *Survey Design Standard 1.2:* Agencies must develop a survey design, including defining the target population, designing the sampling plan, specifying the data collection instrument and methods, ... and be able to measure estimation error.... Documentation of each of these activities and resulting decisions must be maintained in the project files for use in documentation (see Standards 7.3 and 7.4).
>
> ...
>
> *Evaluation Standard 3.5:* Agencies must evaluate the quality of the data and make the evaluation public (through technical notes and documentation included in reports of results or through a separate report) to allow

---

[10] https://www.whitehouse.gov/wp-content/uploads/2021/04/standards_stat_surveys.pdf.

users to interpret results of analyses, and to help designers of recurring surveys.
...

Survey Documentation Standard 7.3: Agencies must produce survey documentation that includes those materials necessary to understand how to properly analyze data from each survey, as well as the information necessary to replicate[11] and evaluate each survey's results (See also Standard 1.2). Survey documentation must be readily accessible to users unless it is necessary to restrict access to protect confidentiality.

OMB (2006) continues with extensive guidance for statistical agencies' transparency efforts in regard to survey data collections, as summarized in Table 1-1.

Statistical agencies have also developed requirements for their internal use to support transparency. An example is the **Census Bureau Statistical Quality Standards,**[12] which contains the following requirements, as summarized in Table 1-2.

Nongovernmental organizations have promoted techniques for enhancing transparency as well. The AAPOR Transparency Initiative,[13] introduced by the American Association for Public Opinion Research (AAPOR) in 2009 and officially launched in 2014, is an effort to help bring about a greater degree of openness in survey practice. The idea behind it is to commend organizations that pledge to practice transparency in their reporting of survey-based findings. AAPOR specifies the following as characteristics of transparency (AAPOR, 2015, pp. 2–3):

Item 2: The exact wording and presentation of questions and response options whose results are reported...;
Item 3: A definition of the population under study and its geographic location;
...
Item 8: A description of the sample design, giving a clear indication of the method by which the respondents were selected, recruited, intercepted or otherwise contacted or encountered, along with any eligibility requirements and/or oversampling;
Item 9: Method(s) and mode(s) used to administer the survey (e.g., CATI, CAPI, ACASI, IVR, mail survey, web survey) and the language(s) offered;
Item 10: Sample sizes (by sampling frame if more than one was used) and a discussion of the precision of the findings. For probability samples, the estimates of sampling error will be reported, and the discussion will state

---

[11]The term "replicate" as used here is likely a synonym for the term "reproduce" as used in this report.
[12]https://www.census.gov/about/policies/quality/standards.html.
[13]https://www.aapor.org/Transparency_Initiative.htm.

TABLE 1-1 OMB Standards and Guidelines for Statistical Surveys: Sections 7.3 and 7.4

OMB Standard/ Guideline	Documentation Required of Agency
Standard 7.3: Survey Documentation	Standard 7.3: Agencies must produce survey documentation that includes those materials necessary to understand how to properly analyze data from each survey, as well as the information necessary to replicate and evaluate each survey's results (See also Standard 1.2). Survey documentation must be readily accessible to users unless it is necessary to restrict access to protect confidentiality.
Guideline 7.3.1: Survey system documentation includes all information necessary to analyze the data properly.	1. OMB Information Collection Request package; 2. Description of variables used to uniquely identify records in the data file; 3. Description of the sample design, including strata and sampling unit identifiers to be used for analysis; 4. Final instrument(s) or a facsimile thereof for surveys conducted through a computer-assisted telephone interview (CATI) or computer-assisted personal interview (CAPI) or Web instrument that includes the following: All items in the instrument (e.g., questions, check items, and help screens); Items extracted from other data files to prefill the instrument (e.g., dependent data from a prior round of interviewing); and Items that are input to the post data collection processing steps (e.g., output of an automated instrument); 5. Definitions of all variables, including all modifications; 6. Data file layout; 7. Descriptions of constructed variables on the data file that are computed from responses to other variables on the file; 8. Unweighted frequency counts; 9. Description of sample weights, including adjustments for nonresponse and benchmarking and how to apply them; 10. Description of how to calculate variance estimates appropriate for the survey design; 11. Description of all editing and imputation methods applied to the data (including evaluations of the methods) and how to remove imputed values from the data; 12. Descriptions of known data anomalies and corrective actions; 13. Description of the magnitude of sampling error associated with the survey; 14. Description of the sources of nonsampling error associated with the survey (e.g., coverage, measurement) and evaluations of these errors; 15. Comparisons with independent sources, if available; 16. Overall unit response rates (weighted and unweighted) and nonresponse bias analyses (if applicable); and 17. Item response rates and nonresponse bias analyses, (if applicable).

continued

TABLE 1-1 Continued

OMB Standard/ Guideline	Documentation Required of Agency
Guideline 7.3.2: To ensure that a survey can be replicated[1] and evaluated, the agency's internal archived portion of the survey system documentation, at a minimum, must include the following:	1. Survey planning and design decisions, including the OMB Information Collection Request package; 2. Field test design and results; 3. Selected sample; 4. Sampling frame; 5. Justifications for the items on the survey instrument, including why the final items were selected; 6. All instructions to respondents and/or interviewers either about how to properly respond to a survey item or how to properly present a survey item; 7. Description of the data collection methodology; 8. Sampling plan and justifications, including any deviations from the plan; 9. Data processing plan specifications and justifications; 10. Final weighting plan specifications, including calculations for how the final weights were derived, and justifications; 11. Final imputation plan specifications and justifications; 12. Data editing plan specifications and justifications; 13. Evaluation reports; 14. Descriptions of models used for indirect estimates and projections; 15. Analysis plans; 16. Time schedule for revised data; and 17. Documentation made publicly available in conjunction with the release of data.
Guideline 7.3.3:	For recurring surveys, produce a periodic evaluation report, such as a methodology report, that itemizes all sources of identified error. Where possible, provide estimates or bounds on the magnitudes of these errors; discuss the total error model for the survey; and assess the survey in terms of this model.
Guideline 7.3.4:	Retain all survey documentation according to appropriate Federal records disposition and archival policy.
Standard 7.4: Documentation and Release of Public-Use Microdata	Agencies that release microdata to the public must include documentation clearly describing how the information is constructed and provide the metadata necessary for users to access and manipulate the data (See also Standard 1.2). Public-use microdata documentation and metadata must be readily accessible to users.
Guideline 7.4.1:	Provide complete documentation for all data files.
Guideline 7.4.2:	Provide a file description and record layout for each file. All variables must be clearly identified and described.
Guideline 7.4.3:	Make all microdata products and documentation accessible by users with generally available software.

---

[1]The term "replicate" as used here is likely a synonym for the term "reproduce" as used in this report.

INTRODUCTION 29

TABLE 1-1 Continued

OMB Standard/ Guideline	Documentation Required of Agency
Guideline 7.4.4:	Clearly identify all imputed values on the data file.
Guideline 7.4.5:	Release public-use microdata as soon as practicable to ensure timely availability for data users.
Guideline 7.4.6:	Retain all microdata products and documentation according to appropriate Federal records disposition and archival policy. Archive data with the National Archives and Records Administration and other data archives, as appropriate, so that data are available for historical research in future years.

SOURCE: U.S. Office of Management and Budget, 2006.

TABLE 1-2 U.S. Census Bureau's Statistical Quality Standard F2: Providing Documentation to Support Transparency in Information Products

Statistical Quality Standard	Documentation
Requirement F2-1	Documentation that would breach the confidentiality of protected information or administratively restricted information or that would violate data-use agreements with other agencies must not be released.
Requirement F2-2	Documentation must be readily accessible in sufficient detail to allow qualified users to understand and analyze the information and to reproduce (within the constraints of confidentiality requirements) and evaluate the results.
Requirement F2-2.1	Descriptions of the data program must be readily accessible.
Requirement F2-2.2	Descriptions of the concepts, variables, and classifications that underlie the data must be readily accessible.
Requirement F2-2.3	Descriptions of the methodology, including the methods used to collect and process the data and to produce estimates, must be readily accessible.
Requirement F2-2.3.1	Measures and indicators of the quality of the data must be readily accessible.
Requirement F2-2.3.2	The methodology and results of evaluations or studies of the quality of the data must be readily accessible.
Requirement F2-2.4	Documentation of public-use data files must be readily accessible in sufficient detail to allow a qualified user to understand and work with the files.

SOURCE: U.S. Census Bureau, 2015.

whether or not the reported margins of sampling error or statistical analyses have been adjusted for the design effect due to weighting, clustering, or other factors;

Item 11: A description of how the weights were calculated, including the variables used and the sources of weighting parameters, if weighted estimates are reported.

SOME CONSTRAINTS

Despite the considerable benefits to being transparent, it is important to stress that the answer is not simply, "More is always better." More transparency may not be legal or feasible. For example, the release of some input datasets from surveys could make confidential personal information for individuals or businesses available, which would violate the law as well as undermine trust in the statistical agency. Further, because statistical agencies are increasingly relying on nonsurvey data, including administrative, commercial, and other digital trace data, the data may come with additional restrictions on transparency. There are laws and interagency memoranda of understanding that often prevent the disclosure of administrative data. Commercial data or contractor intellectual property may be protected by contract for commercial reasons. The same is true for some digital trace data. In some cases, methods for producing estimates cannot be shared because doing so could make the estimates susceptible to manipulation (e.g., release of the stores that surveyors visit to record prices to estimate the Consumer Price Index).

There are resource costs associated with achieving greater transparency, so investments in transparency should meet a cost-benefit test. In some cases, where there is limited interest or benefit from a more detailed level of technical documentation, the benefits might not justify the costs, especially for perennially resource-constrained statistical agencies. Cost-benefit tradeoffs are discussed further in Chapter 2.

More generally, not all users look at statistical estimates with the same expectations or having the same fundamental knowledge of statistics. Transparency requires documentation that is appropriate and accessible to the user: excessively technical documentation may not make statistical agency products more transparent to some users. Some users may just want to know the quality of the estimates and how the estimates can be safely used but will not want to wade through complicated software code in order to understand what underlies the methodology used. Other users may only want to have an overview of how the data were treated and the estimates produced. In contrast, academic researchers may wish to have various methodological details that can only be examined through access to (a subset of) the complete (commented) code. Statistical agencies need to

INTRODUCTION

conduct user experience testing and produce information at multiple levels for different users with different needs.

Transparency should not be considered as just a passive interaction. Rather, it should be understood as an active effort, a way of making information available to a user community so its members can know how a set of official statistics were produced and what their resulting quality is. In this communication, both the audience (the user community) and the matter being communicated are important, with different users needing different portions of the details provided.

REPORT STRUCTURE

As noted above, this report has a dual focus on both NCSES and the overall federal statistical community. Recommendations are offered to NCSES, along with recommendations that apply to all the major federal statistical agencies, on how to embrace the value of transparency.

Chapter 2 starts with a description of the complexity of the federal statistical programs and an explanation as to why documentation is a challenge. It details the various advantages that stem from greater transparency in the production of official statistics. The chapter then describes existing requirements, followed by our assessment of existing practices regarding documentation and archiving, including what OMB requires, records schedules and data management plans, and documentation of data treatment techniques and estimation methodologies. It concludes with a discussion of constraints and arguments for less than complete transparency in some situations.

Chapter 3 examines changes in archiving practices that can improve transparency and reproducibility, including the role of the National Archives and Records Administration and the implications of the FAIR (Findable, Accessible, Interoperable, and Reusable) principles[14] and existing OMB directives. The chapter addresses the role of catalogs and searchable metadata repositories in achieving "findability" and the importance of paradata for scientific innovation.

Chapter 4 discusses changes in documentation practices that could facilitate transparency and reproducibility of the statistical methods used in the federal statistical agencies. These includes various information technology tools to assist with version control and software development, as well as collaboration tools and methods for retaining workflows. Also discussed is the necessity of improved interactions with the public.

Chapter 5 examines the contribution that metadata standards make to transparency and reproducibility. It defines metadata and discusses the risks

---

[14] https://www.go-fair.org/fair-principles/.

and benefits of using metadata systems for documentation and archiving and how such systems could be integrated into the current systems in use at the agencies. The chapter provides a summary description of common metadata standards currently in use and discusses how increased use would affect transparency.

Chapter 6 presents best practices for NCSES to enhance its transparency. It describes NCSES's current programs and their existing publication standards, some areas for improvement, and recommendations as to how such improvements could be arrived at.

Finally, Chapter 7 provides a discussion of best practices for federal statistical agencies. This chapter lays out aspects of what a more modern approach to federal statistics would consist of, including better documentation and archiving practices that provide for data and methodology sharing among agencies and a smoother interface with the public. Recommendations as to how this new vision can be initiated are provided.

2

Current Practices for Documentation and Archiving in the Federal Statistical System

THE COMPLEXITY AND SCIENTIFIC NATURE OF THE PRODUCTION OF OFFICIAL STATISTICS

The production of official statistics is more complicated than many users of federal statistics understand. It requires the collection and quality control of data from multiple sources, and it often uses models to produce valid estimates and to protect confidentiality. When statistics are based on collected survey data, their production can involve survey design (primary sampling units, stratification, clustering), design of the survey instrument, field instructions, various data treatments to prepare the raw data for input into the estimation program(s), an estimation methodology, and whatever validation is used to assess the quality of the resulting official statistics. If estimates instead use data from administrative records or digital trace data,[1] other technical questions may have to be dealt with.

More involved yet is the increasingly common situation in which survey data, administrative data, and digital trace data are used in some combination to exploit their inherent strengths and minimize their weaknesses. This can involve the use of models that need to be fitted and assessed. Whether the input data come from a survey, administrative records, or digital traces, each of these steps may require addressing various complex questions about how to collect the highest quality input data, or what methods will

---

[1] Unfortunately, there is no commonly accepted term for the new sources of data that technology has made possible, which includes *Internet transaction data, social media data,* and *sensor data.* Other terms that have been suggested for (some of) this new type of data are *organic data* and *big data.*

best treat the raw data for nonresponse or edit failures, or what model will produce the highest quality estimates. So, each stage—the collection of the input data, the processing of the input data, the computation of the estimates or indicators, the production of the associated metadata, and the evaluation of the resulting official statistics—can raise difficult issues that need careful resolution. And many of these steps can involve difficult scientific questions, some of which may have been addressed in the technical literature, and some of which may be novel and require innovative thinking.

To illustrate these issues, consider a survey conducted by the National Center for Science and Engineering Statistics (NCSES), the National Survey of College Graduates. In 2017, this survey, which is conducted every other year, sampled some 124,000 graduates, drawn from individuals who previously responded to the American Community Survey. The design of the study is a rotating panel: respondents are asked to respond to the survey four times, at 2-year intervals over a period of 8 years. Each survey year, a new cohort is recruited to serve in the panel, as the oldest cohort rotates out of the sample. Thus, each cross-sectional sample consists of new respondents and others who have responded in earlier waves. The rotating panel design is intended to capture changes in education and occupation circumstances in the U.S. adult population while limiting the participation burden on each respondent. An extensive questionnaire inquiring into educational experiences and their relation to occupational outcomes, particularly in the sciences and engineering, is administered in the first year of participation, with follow-up questionnaires completed in subsequent years. Three modes of data collection are employed to gather responses: Internet questionnaire, mail questionnaire, and telephone interview. Generally, the less expensive Internet or mail modes are tried first, followed by telephone if responses are not obtained with those self-administered methods. For the panelists in follow-up cohorts, the mode employed may be adjusted to use the one that achieved a response in the first wave. Unconditional incentive payments to encourage participation are made to some respondents whose response propensity is estimated to be lower and whose base weight (a measure of rarity in the sample) is high.

There are a number of post-survey adjustments made to the resulting sample. Every sample case has a weight that is a product of sample selection (base weight), weighting to account for unit nonresponse, trimming to eliminate extreme weights, raking procedures to bring sampling weights in line with sample frame estimates, and procedures to convert weights calculated for each sampling frame from which respondents are drawn into a final weight that reflects the population for the survey year. There is also use of imputation employed for item nonresponse. Many users of official statistics have an incomplete understanding not only of the complexity of the collection of the raw input data, but of the measures taken to turn those input data into the final official estimates.

Simply put, official statistics are scientific estimates, which can sometimes be relatively complicated. This is another reason why a greater degree of transparency is important to have.

The committee's impression is that the documentation of methods, the archiving of collected data, and the archiving of the resulting estimates in the federal statistical system are often viewed as having a low priority, and therefore that these actions are often carried out as an afterthought or not at all, depending on various time pressures. We strongly urge, instead, that these activities be treated as an integral part of the process of the development of a set of official statistics. They should be planned and accomplished alongside the work of constructing official estimates.

WHY TRANSPARENCY AND REPRODUCIBILITY ARE GOALS FOR NCSES AND THE FEDERAL STATISTICS SYSTEM

The benefits of transparency and reproducibility can be broken down into four pieces: *efficiency, innovation and progress, trust and confidence*, and the *value from the use of the data products themselves*. These four dimensions are now described.

Efficiency

As mentioned, it is important to know how an agency produces a set of official statistics, since that informs one about the quality of the methods used, and therefore the quality of the resulting official statistics, and it provides proof that no special interests are playing any role. Knowing how a set of official statistics is produced entails retaining details of the various data collection processes used, including the survey design, the survey instruments and field instructions, should survey data be collected, or retaining descriptions of any processes used to collect data from administrative sources and/or from digital traces. Then, whatever computations are carried out in converting the raw input data to the set of inputs fed into the estimation methodology also need to be documented. This is done by retaining the commented code used in the workflow that takes the set of raw data, modifies it for nonresponse, failed edits, and other corrective actions, and carries out all data transformations needed. Finally, the commented code that describes how the estimation procedures are applied also needs to be retained along with the relevant computing environment, as well as the computations carried out to assess the variability of the official statistics and efforts to validate them.

In all, as part of any organization that is involved with such complex processes, it is necessary for statistical agencies to retain, in an accessible way, detailed information as to how they accomplish their various tasks in the production of a set of official statistics. By doing this, these agencies can accommodate changes to staff as discussed in Chapter 1. Also, such a careful description makes clear to staff that their work is subject to detailed review, and therefore encourages thoroughness and care. This reflects a practical business case for transparency.

In addition, as part of such detailed documentation, when changes are made it is important not only to make the associated changes in the documentation, but to make available an explanation as to why the changes needed to be made. The precise location of the changes in the software code, either for the workflow history of the data treatments or for estimations, also needs to be made available, depending on what has changed. That way, if the changes are later found to be the cause of any problems, it will be a simple matter to undo them, and possibly to understand where the logic behind the change went wrong. Many users are concerned when changes in procedures or estimation methods are made, since such changes can then be responsible for changes in the official statistics that are not due to real changes in the phenomena being measured. Therefore, it is important to collect such changes in a specific, permanent location online so that users looking at time series of official statistics can possibly determine when changes in the series are (and when they are not) due to such modifications.

Innovation and Progress

Science marches on, and the science underlying federal statistics is no different. It is very likely that we are currently in a period in which substantial changes are either occurring or will soon occur as a result of the greater cost of collecting survey data and the greater use of other sources for input data. There are many reasons for innovation and progress, including the dynamic nature of the population and businesses whose characteristics we need to measure, opportunities provided by advances in technology, and theoretical progress. Changes in society currently include a greater reluctance to participate in surveys. Technological changes include the growth of mixed survey-mode designs, which are jointly motivated by a greater reluctance to participate in surveys generally and the rising cost of the traditional face-to-face contact method. The mixed-mode designs often involve attempts to get responses by mail or through self-administered questionnaires on the Web, followed by more expensive telephone or face-to-face contacts for nonrespondents. These changes in design are accompanied by new software tools that facilitate collaboration in software development, by making software code more easily available to the public, and the development of standardized metadata, all of which are discussed in Chapters 4 and 5.

Examples of new methods that have resulted from theoretical developments include the application of model-assisted survey sampling (including generalized regression estimators and calibration estimators), the application of Fay-Herriot models, often to combine information from disparate sources, and sample reuse methods for variance estimation. These improvements and innovations can arise either within an agency or through the external research community, and therefore both internal staff and external researchers need to be aware of the current approaches used for the production of a set of official statistics so that they can assess when a new idea might provide an improvement over the status quo.

Trust and Confidence

If statistical agencies wish the public to trust the estimates coming from the government, they should recognize that it is important to demonstrate to the public that the techniques used for data collection and the methods used on the collected data are not used to benefit any stakeholders and that they represent the current state of the art. This is accomplished by making their official statistics, their data collection techniques, and the estimation methods used to produce them available to the public, to the extent possible. Further, making available the data collection techniques and computations, along with the relevant input datasets—under secure arrangements to protect confidentiality—permits the validation of a set of official statistics by demonstrating its computational reproducibility.

Value from the Use of the Data Products

If a data product is not fully documented, for example when an agency does not provide the weights necessary to support valid inferences (if the data are survey based with nonresponse) or does not provide the sampling variances (in cases where there is interest in combining estimates through use of a statistical model), then users are likely to produce substandard estimates or analyze the data inappropriately. As described in detail in Chapter 5, fully documented resources (e.g., datasets) are realized by providing metadata that conform to some predefined metadata schema, preferably a publicly accessible standard. The ultimate goal is to develop fully documented datasets through use of machine-actionable, organized, comparable, quality metadata tied to the relevant data and making all of it accessible through an application programming interface.

If agencies exchange such documented information, users are able to resolve questions of the following nature: whether it is appropriate to utilize a tabulation that would require 10,000 cells, or what weights to provide for the survey-based estimates used in combination with a set of model-based estimates using administrative records in a small-area estimation model. For time-series analyses, the metadata could provide information on major changes to the series, which would inform the user about whether indicator variables for those changes should be incorporated in such models.

Further, more generally, use of algorithms developed for the production of one data product may become useful for the production of another. Standardization of measurement for common concepts, where appropriate, could become possible. Understanding of the differences in estimates of phenomena measured in different surveys could be enhanced. Agencies would be better able to track the impact of changes in collection methods. Continuing surveys or other collections would be more easily and accurately carried out.

In sum, by enabling users to have a better understanding of the characteristics of a set of official statistics, agencies enable those statistics to gain in value by being put to use for additional purposes.

EXISTING REQUIREMENTS

The U.S. Office of Management and Budget (OMB) provides guidance to the agencies concerning what data collection methods (mainly for surveys) federal statistical programs are obligated to make public. Specifically, OMB's *Standards and Guidelines for Statistical Surveys* (OMB, 2006) describes the process agencies must follow to receive permission from OMB to collect survey information. Once a survey is approved, OMB makes its information collection review publicly available at https://www.reginfo.gov/. Two parts of this OMB product, sections 7.3 and 7.4, directly address documentation of data collection methods.[2] It is the case that the federal statistical agencies do comply with this request for information from OMB since OMB's approval is a requisite to initiate data collection.

These standards say much less about statistical programs that are based on administrative data or digital trace data. Similarly, OMB provides less guidance with respect to the documentation of methodologies either used for data treatment or for estimation associated with the official statistics, including whether the associated software code should be made known to the public. (While these standards do not specify much regarding documentation of data treatment, some summary measures on data quality relative to nonresponse are required.)

With respect to guidance about archiving the input data and the official statistics, records schedules do exist informing agencies what is to be

---

[2] It should be mentioned that sometimes changes are made after approval is granted, but in those instances OMB requires that a change request be made or, if substantive enough, approval of a new submission. In any case, revised plans are ultimately required by OMB within 3 years and therefore what is made public will soon be consistent with the survey data collection processes used.

retained at the National Archives. However, more comprehensive directives, often found in data management plans, about what precisely is to be retained, where it is to be located, for how long, using metadata from which standard, and whether the public is to have access to it, are currently not produced.

Most federal statistical agencies have developed their own statistical standards and reporting guidelines in support of greater transparency in their programs. For example, the U.S. Census Bureau's "Statistical Quality Standard F2—Providing Documentation to Support Transparency in Information Products" identifies topics that should be discussed or described in a survey's technical documentation (see Chapter 1). Similarly, NCES has developed statistical and reporting standards for data collection and statistical processing. NCES also stipulates standards for the dissemination of its data, including which topics should be reported in a statistical program's technical documentation.

The approaches used by the statistical agencies to establish statistical and reporting standards vary—some agencies being very prescriptive and others less so. Also, we have not examined the extent to which these standards or guidelines at the individual agency level are implemented in practice.[3]

EXISTING PRACTICES

In most agencies, decisions on what to retain and what to release are made at the programmatic level. Agency standards, if they exist, obviously provide only a minimum standard, which many programs decide to exceed. For example, at the 2017 workshop on transparency (NASEM, 2019a), representatives of the Longitudinal Employer-Household Dynamics program at the Census Bureau reported that they have the capability of recovering any input data set, the associated program code, and the resulting official estimates in a few minutes.

What documentation is available very likely depends on a variety of things, including the perceived importance of the program, whether the program is a one-time program providing estimates for a single occasion or a continuing program, the resources made available by the agency's parent organization for the overall study or project, and whether the data collection and/or data treatment was carried out by a contractor, the Census Bureau, or in house by the agency. Also, the funding for documentation can be affected by issues that arise during data collection. For instance,

---

[3] A complete listing of agency reporting guidelines and statistical standards is provided by the Interagency Council on Statistical Policy. A link to this is provided by NCES through the Federal Committee on Statistical Methodology (see https://nces.ed.gov/FCSM/policies.asp).

additional resources might be needed to obtain enough sample items, or the price of external data might increase. Agencies will often prioritize data collection and, as a result, take funds from reporting budgets. Also, agencies are often required to conduct the same data collection operation over time with a fixed budget, resulting in a similar problem of having to decide whether to collect more data or provide documentation.

In order to learn what was going on at the program level, we communicated with a small number of high-profile statistical programs by means of an informal questionnaire to see what documentation and archiving practices they used. We assumed that for lower-profile programs and for single-use circumstances the documentation and archiving practices were likely to be less complete. We sent this informal questionnaire to the program managers or high-level staff for 20 high-profile programs, and we received 11 responses, all from the programs listed in Box 2-1. We asked these individuals about their practices regarding the documentation of their data collections, the archiving of the resulting input datasets and resulting official statistics, and the documentation of statistical methods used to treat the data and to produce the indicated official statistics.

RESPONSES TO THE INFORMAL QUESTIONNAIRE

By requesting that program chiefs or other informed staff respond to these informal questionnaires, the committee was able to get a rough sense of what the current practice is, both internally and externally, regarding the documentation of the data collection methods, the data treatments used,

---

BOX 2-1
Programs That Responded to Informal Panel Questionnaire

1. American Community Survey, Census Bureau
2. National Marine Fisheries Service, National Oceanic and Atmospheric Administration
3. Residential Energy Consumption Survey, Energy Information Administration
4. Food Security Survey Module, Economic Research Service
5. National Crime Victimization Survey, Bureau of Justice Statistics
6. Statistics of Income Tax Data, Internal Revenue Service
7. Consumer Price Index, Bureau of Labor Statistics
8. Disability statistics from the Office of Research, Evaluation, and Statistics, Social Security Administration
9. Population Estimates Program, Census Bureau
10. National Health Interview Survey, National Center for Health Statistics
11. American Time Use Survey, Bureau of Labor Statistics

and the estimation routines employed. In addition, this informal questionnaire asked about the current policies regarding assessment of input data quality, where the input datasets and the resulting official estimates are stored, with what additional information about the data file, and for how long they are to be retained.

Internal Retention Policies

The questionnaire began by asking about *internal* retention policies, starting with whether input datasets were saved in some type of repository internal to the agency and, if so, for how long they were retained. Other questions concerned whether the program was a continuing program, whether every new data collection was saved in its own location, and whether metadata were provided along with the saved data files (and, if so, whether a metadata standard was used). Most respondents said that the input datasets were saved on internal directories and that they were generally not overwritten by subsequent data collections. Further, while these files were stored with relevant metadata, they said that typically no metadata standard was used.

Respondents said that care was taken to include everything needed to understand what the file contained and to support later reuse. The input data files retained were generally the edited files used as input to produce the associated official estimates, they said, but some programs retained the raw data files and other intermediate files prior to the production of the final input data. There often were internal guidelines for what and how to save but, respondents said, there were typically no guidelines on the use of metadata standards.[4] Some agencies pointed out that there was no repository on their own (internal) Web page that stored all the data intended for later reuse. One agency used the Inter-University Consortium for Political and Social Research (ICPSR) as its archive, and other agencies used the National Archives and Records Administration to archive data.

Retention of Workflow History

The questionnaire next asked whether the workflow history of data treatments was retained somewhere, possibly as commented code, and whether the estimation methodology was likewise retained, again as commented code. It also asked whether there were technical memoranda that summarized the approaches used for these computations. Here the responses were fairly consistent. Commented code was retained internally and

---

[4] An exception is the use of International Organization for Standardization standards—see https://www.iso.org/home.html.

stored in a way that supported later reuse, certainly in support of reproducibility studies with the above input datasets, but also for general research purposes. In addition, the agencies often made available a user's guide and technical memoranda which were written to be accessible to a broad user community. Some respondents pointed out, however, that the comments and memoranda focused on what had been done with little discussion of how or why.

Some programs made extensive use of administrative records as input data. In those cases, it was typical for the disaggregated administrative records not to be released in any fashion, even in a restricted sense to a federal statistical research data center (FSRDC).

Use of Metadata Standards

The next two questions asked whether the Data Documentation Initiative (DDI), Statistical Data and Metadata Exchange (SDMX), or other metadata standards were used and whether the decision on whether to use such standards was based on a cost-benefit assessment of doing so. There were two mentions of the current use of DDI, and a future intended use of DDI (one of these was so that the data could be archived and disseminated at the ICPSR), and there was one mention of SDMX. Otherwise, the adoption of these standards or other metadata standards seems not to be typical of most federal statistical agencies.

One agency expressed concern that to adopt these standards would obligate the agency to do so for the entire historical series to ensure backward comparability. This is something that many agencies could be faced with. Our view is that backward compatibility should be achieved over time, but the focus should be on moving newer data to more open standards. We also note that when standards are open, contributions of this kind can also come from the user community. This is the approach that the Consumer Expenditure Surveys are taking.

What Is Made Available to the Public

The questionnaire asked what was made available *externally* to the public. We asked whether input datasets were available in FSRDCs, and if so what metadata were provided with the data and the standards that these metadata satisfied. Also, use of any alternatives to an FSRDC were explored, especially use of public-use microdata sets. The release of public-use microdata sets was quite common, and these were often accompanied by user's manuals indicating how to use such files for analytic purposes. Some agencies also make arrangements for the release of microdata to researchers in other secure environments.

Documentation of the survey design and the wording used in the survey instrument was always available, but sometimes only by request. Typically, considerable information was also available on the workflow history of the data treatments and the estimation methodology used. It was not uncommon for the commented code for either the workflow history or the estimation methodology to be made available as well. (There were even efforts to ensure that an external user could check on the computational reproducibility of some official estimates.) However, while there were some outstanding examples of accessible descriptions of the data treatments and estimations—for instance, the *American Community Survey Design and Methodology (January 2014)*—it is common for no such documents to be available.[5]

Concerning documentation of data quality, many agencies provide a coefficient of variation for the official statistics (when they are weighted aggregates from a survey), as well as the frequency of nonresponse. Many agencies conduct research on issues such as the data treatments used and variance estimation, and commonly make the results available as technical reports or white papers, sometimes published in the technical literature. Four journals in which such papers appear are *Survey Methodology*, the *Journal of Official Statistics*, the *Journal of Survey Statistics and Methodology*, and *Public Opinion Quarterly*.

Limitations of This Questionnaire

There are many limitations that must be conceded with this informal information gathering. To begin with, it was a very small "sample" with considerable nonresponse. Further, some questions requested detailed responses but respondents sometimes struggled to provide all the processes and details the questionnaire was meant to elicit. The wording of some questions often made implicit assumptions about the data collection or the methods that made the questions difficult for respondents to interpret in

---

[5] The federal statistical agencies have written a number of excellent technical reports and handbooks for various official estimates that are based on the data collected from surveys. A key example of this is U.S. Census Bureau's *Current Population Survey Technical Paper (CPS TP77)* (U.S. Census Bureau, 2019). Also, there have been efforts to provide contextual documentation for public release datasets, including documentation of the Public Use Microdata Sample Files for surveys such as the American Community Survey (https://www2.census.gov/programs-surveys/acs/tech_docs/pums/ACS2014_2018_PUMS_README.pdf). In addition, the statistical agencies have funded data quality assessments for surveys such as the Survey of Income and Program Participation (SIPP) and for the American Housing Survey (AHS)—for the SIPP, see Kalton (1998) and Jabine, King, and Petroni (1990). For AHS see Chakrabarty and Torres (1996). Further, the agencies have regularly published articles in the scientific literature informing those interested in the technical details of the methodology employed in various official estimates. Examples include Fay (1984); Dippo, Fay, and Morganstein (1984); Findley (2005); and Zieschang (1990).

some situations. For example, the questions were somewhat survey-centric and did not provide enough guidance to programs that used extensive nonsurvey data. Finally, we asked about machine-readable metadata standards only with regard to nonsurvey data, which was an error.

Given these caveats, in combination with our knowledge of what OMB requires, the existence of several existing guidelines about documentation—in particular that of the Census Bureau—our examination of the Web pages for the same programs, and finally our direct knowledge of the documentation that certain programs release, we believe we have a reasonable idea of the range of documentation that is currently provided both internally and externally to an agency.

IMPLICATIONS OF INFORMAL QUESTIONNAIRE RESULTS

With respect to internal saving of their input data, all agencies retain these data and the official estimates for at least a year. Some agencies keep such data for as long as 40 or more years. (The Census Bureau keeps data for much longer than that.) Some agencies save many versions of the input data, including the raw responses and the file containing the responses treated for edit failures and for nonresponse; the most commonly saved internal version is probably the file used as final input to generate the official estimates.

With respect to the metadata provided with the saved input data files, many data programs use SAS files to indicate variable definitions and produce codebooks to indicate variable locations on the files. Two data programs use DDI as their metadata standard, another is working toward implementing it, and one uses SDMX. No programs try to evaluate whether to adopt metadata standards through a cost-benefit analysis. Further, no programs try to indicate the quality of their administrative record data by providing something like the nonresponse rate or the rate of undercoverage.

The workflow history is generally available internally with commented code. The same is true of the estimation methodology. Therefore, the statistical programs can typically support computational reproducibility internally, at least for a short period of time.

Regarding access to input data for external users, given that it is almost always confidential, the only access external users will have to the disaggregated input survey data is inside an FSRDC or other secure computing environment. Some agencies make available public-use microdata samples, and some metadata are saved with the input data, generally in the form of codebooks and other agency-specific procedures.

Regarding information on data collection methods, data treatments, and estimation methodologies provided to external users, there is often substantial information on the survey design used and the survey instrument.

This is unsurprising, since most of this information is also available in the information collection requests submitted to OMB for data collection clearance. In addition, considerable information is available on the workflow history and the estimation used in technical methods documents, handbooks, white papers, and the statistical research literature. Surprisingly, even the commented code associated with these methods is at times made available to the user community.

A last issue, though not addressed in our informal request for information, is the need to help the external user community recognize when the processes used for data collection have changed substantially—for example, when variables have different content, or they have different locations on the data file, or when tables for different cross-tabulations have changed. Having a location on each program's Web page that indicates all the major changes that have been implemented since the previous data collection would be very useful. This is sometimes done but not uniformly.

In general, agencies provide guidance that appears to be survey-centric. They focus on details of the data collection and not on details of the data treatment or the estimation methodology. They provide little information on the extent to which the data are saved or archived, where they are saved or archived, and how to gain access to such data. They rarely use community-normed metadata standards.

There is considerable variability across programs, and what is available for lower-priority programs is likely to be less comprehensive than what is available for ongoing, highly visible programs. There is probably less transparency about the collection and use of administrative data or digital trace data for input into official statistics programs than for survey inputs. Agencies provide little guidance to their programs about saving or archiving (preserving) data, especially input data but also public, official statistics. They also provide limited guidance about the kind of metadata and paradata that should be produced and shared.

Finally, transparency requires accessibility, so it is important to determine the difficulty or ease in finding technical documentation. To check this, we looked to see how far removed these documents were from the home page of each program. In most cases, there was a link on the program's home page that referred to "Technical Documentation" (or that used a similar phrase), so these documents were often only a single click away from the landing page.

Conclusion 2.1: Documentation of data collection methods, data treatments, and estimation methods by federal statistical agencies, while in need of some improvement, is generally fairly complete with respect to what is available internally to an agency. The practice of archiving input datasets and official estimates varies greatly across agencies, and as a result some data are not retained even internally for long periods of time. Externally, while the public sometimes can gain access even to the code for various methodological processes, agencies often do not provide accessible methodological summaries for nonspecialists. Further, access to input datasets using secure avenues varies substantially across agencies.

CHALLENGES THAT ARISE IN IMPLEMENTING TRANSPARENCY AND REPRODUCIBILITY

There are challenges and costs associated with the use of increased transparency. To start, there is generally widespread agreement, though not universal practice, in agencies creating comprehensive, detailed documentation of all the steps used in the production of a set of official statistics for all the reasons given at the start of this chapter. Therefore, the costs of transparency due to what is needed internally are unavoidable. What is less agreed upon is what should be made publicly available. It is not the case that what is made public is always or should always be a subset of such internal documentation, since it is often desirable to provide information in a less technical and less detailed form, given the needs of many types of users. As a result, there may be added costs to providing external documentation to the public. It needs to be emphasized that the following challenges to complete disclosure of data and methods only affect the pace of movement toward full disclosure, but not the ultimate goal of full transparency.

Resources

Resources is our term for the time it takes staff to prepare various documents and materials needed to inform the public about how a set of official statistics was produced. In addition to time, there may also be an opportunity cost, since the people best suited to produce at least some of these documents and materials are likely to be the most talented staff. These are people who could be devoting their scarce time to the refinement of methodologies and the development of other methodologies. At the same time, agencies realize that the subset of their user community that is interested in details concerning the data collection or the various computations undertaken is likely quite small. Therefore, it is not unreasonable for agencies to decide that the scarce resource of their technical staff members' time could at times outweigh the needs of a small fraction of their user community.

The resources needed will also include the time spent on emails and on the phone with expert users who wish to find out precisely what was done and why, in response to this additional degree of clarity (though the argument can be made that in a world where good transparent documentation is available, the number of calls from users is likely to be substantially reduced.) But it is conceivable that the desire for additional transparency, especially for nontechnical users, might lose out to resource constraints.

Participant/Respondent Confidentiality

Input datasets often have personally identifiable information (PII), which is forbidden by law to be released to the general public. As a prime example, data from individuals on federal household surveys include PII and therefore are not shareable unless care is taken to virtually eliminate the chances of a disclosure. Also, if federal statistical agencies use administrative data at the national or state level to develop official statistics, there are often laws forbidding any sharing of such data. In the case of survey input data, often the only way to make such input datasets available for analysis to members of the external research community is to do so in secure environments that provide comprehensive protection against disclosures. Such datasets are therefore only provided through use of federal statistical research data centers and comparable constructs, they are anonymized, and additional techniques, such as differential privacy, are applied to them to reduce any remaining risk of disclosure. Finally, any material the researchers wish to take with them from these federal statistical research data centers is carefully checked to see if there is any risk associated with their doing so. Even so, many types of data are not available even under such protections.

Manipulation of Estimates by Third Parties

The concern here is best communicated using an example. Suppose that there is transparency about how the consumer price index is produced. This price index is used for contract indexation and can rely on responses from a small number of firms, including firms whose contracts are indexed to the measure. These responses could be intentionally mis-responded to in order to artificially manipulate the estimates. Such concerns are not limited to the official statistics produced by the Bureau of Economic Analysis. Any policy-related intervention may need to keep certain parameters secret, at least while the intervention is ongoing. Therefore, if an agency is not careful, efforts to enhance transparency can provide information that can be used to compromise policy assessments or disclose PII.

Transparency and Reproducibility When
Third Parties Collect Data for a Statistical Agency

Complications also arise when multiple agencies work in collaboration or an agency relies on a private-sector contractor. Many agencies rely on the Census Bureau for data collection. The statistical agencies do not receive raw responses if the data are collected under the Census Bureau's Title 13 regulation. Instead, the Census Bureau provides the official estimates that are estimated from the raw data. This limited transparency and reproducibility can undermine the sponsoring agency's ability to improve its own official estimates or serve its users. This situation, we believe, also creates a minor communications problem when information on a set of official statistics has documentation that is split between the statistical agency and the Census Bureau. Further, the official statistics are often transmitted to the agency from the Census Bureau in tabular form, so the agency cannot by itself provide the raw responses to an FSRDC or to a researcher in response to a special request.

Similar issues arise when making use of a private-sector contractor to collect data and/or produce the estimates, as the agency may receive estimates in tabular form, complicating the assessment of data quality. Also, contractors may view the software used for data treatments or for estimation as being proprietary, and as a result may not wish to share this software.

This constraint on transparency is not, however, a necessary outcome. It is important for statistical agencies to realize, first, that openness does not preclude licensing—one can provide code under a noncommercial-use license, and separately contract for commercial use. Second, the agencies control the language that goes into Requests for Proposals, not the contractors. If a (future) contractor wishes not to submit because doing so would reveal what they consider to be proprietary software, then so be it. Third, this is about openness. Even commercial software used in government contracts could be required to be made available, at reasonable prices, to others, which at a minimum would ensure the ability to do "blackbox horse races" or other evaluations. Finally, there is no reason to hide the software code, and it has been standard practice for software companies to have a special license for government use that does *not* make the software become open.

Further, the metadata information that accompanies what agencies get from contractors is often in PDF form, which is often difficult to process. This raises other concerns, because such contractors then serve, in a sense, as the location where the raw data are saved.

Intellectual Property of Contractors and Commercial Entities

As just mentioned, when contractors provide assistance on data treatments or data collection, some of what is carried out may be considered proprietary. This might include a technique for imputation for nonresponse, or some manner of identifying the proper respondent from a business. Commercial entities also can provide datasets at a price that might be used to construct target populations for sampling. In such cases, contractors and commercial vendors may provide their algorithms or data to federal statistical agencies for a specific use for a limited period of time during which they are not to be shared.

In each of these cases where agencies could decide that full transparency is not feasible or is not cost efficient, there remains the obligation to inform the public about what has not been released and the reasons for doing so.

Conclusion 2.2: A foundational element of agencies that produce federal statistics is transparency of operations, methods, and results so that users can trust that federal statistical estimates are produced in an unbiased manner and understand their properties and how best to use them. The principle of transparency is reinforced in numerous reports, directives, and legislation, including the Foundations for Evidence-Based Policymaking Act of 2018 and the Federal Data Strategy.

Recommendation 2.1: Leadership at the Office of Management and Budget, the Interagency Council on Statistical Policy, the National Center for Science and Engineering Statistics, and all agencies that produce federal statistics should establish transparency of processes and methods as a high priority and continuously reinforce this priority to their staffs.

3

Changes in Archiving Practices to Improve Transparency

This chapter explores the relationship between archiving and transparency, and then compares current practices with recommended best practices. Archiving means preserving objects. That implies both selecting which objects will be preserved and which will *not* and taking actions to maintain the possibility of access to those objects that are preserved. Storage alone is not necessarily preservation, and this is especially true in an age in which many objects are "born digital."

The National Archives and Records Administration (NARA) is focused on preservation and therefore is an archive in the classic meaning of the term. NARA takes physical custody of government records and then strives to preserve them, according to stated goals, strategies, and methods for physical[1] and digital[2] records. These records have already been subject to decisions about whether or not they should be archived, in the form of "records (control) schedules." A government agency is legally obligated to "manage" records—that is, to make decisions about creating and preserving such records—in order to document the functioning and decision making of the agency (44 U.S.C. 3101–3107).[3] Records may be "disposed" (destroyed) as well, but such decisions are captured in the records schedule, of which NARA keeps a copy in a searchable archive (https://www.archives.gov/records-mgmt/rcs). NARA is the agency that decides what to

---
[1]https://www.archives.gov/preservation/storage/specs-housing-exhibition-2015-current.html.
[2]https://www.archives.gov/preservation/electronic-records/digital-preservation-strategy.
[3]Records Management by Federal Agencies (44 U.S.C. Chapter 31): https://www.archives.gov/about/laws/fed-agencies.html.

preserve and for how long, subject to guidance[4] and review by the archivists of NARA. NARA provides access to its archives through the National Archives Catalog[5] and its archival databases.[6]

TRANSPARENCY AND ARCHIVES

Archives increase the transparency of federal statistics by providing third parties (or the statistical agencies themselves at another point in time) with the records of what statistical agencies have collected and how they have used what was collected to produce statistical estimates. It is of course equally important that there be transparency prior to and during the data collection process, well before any long-term preservation in an archive (see relevant sections in Chapter 7, Tables 7-1 through 7-6). Archiving also facilitates improvements in statistics, particularly in a system as decentralized as the U.S. federal statistical system. This is because other agencies (and external users) can learn from successful innovations as well as the challenges faced by others, but this learning occurs only if the other agencies can find and analyze what has been done. A transparent policy about what is preserved—and what is not preserved—is important as well for confidence in the statistical system. As Office of Management and Budget (OMB) memo M-12-18[7] states:

> Records are the foundation of open government, supporting the principles of transparency, participation, and collaboration. Well-managed records can be used to assess the impact of programs, to improve business processes, and to share knowledge across the Government. Records protect the rights and interests of people, and hold officials accountable for their actions. Permanent records document our nation's history. (p. 1)

For the federal statistical system, transparency regarding the key elements of both the process and the results of data collection is critical to public confidence. Transparency is thus also critical to public participation in and provision of information *to* the federal statistical system, improving data quality and reducing the costs of data collection. To provide this transparency, it is necessary to archive—to preserve and make accessible—the full data life cycle, including questionnaires, metadata about the data collection process, metadata about the transformation of raw data into data products, and the data products themselves. Archiving allows us to learn from past practices and to put current data into context by understanding

---

[4] https://www.archives.gov/preservation/electronic-records/digital-preservation-guidance.
[5] https://www.archives.gov/research/catalog.
[6] https://aad.archives.gov/aad/.
[7] OMB Memorandum M-12-18, https://www.archives.gov/files/records-mgmt/m-12-18.pdf.

those practices. But more importantly, archiving can serve to enhance trust and confidence in the federal statistical system.

More directly relevant to the question of the transparency of federal statistics is the 2019 OMB memorandum on the Transition to Electronic Records, which requires all federal agencies (not just statistical agencies) to "Ensure that all Federal records are created, retained, and managed in electronic formats, with appropriate metadata."[8]

There are general principles, codified in legislation, regulations, and policies of the United States, as well as the United Nations and the Organisation for Economic Co-operation and Development (OECD), that guide and provide some specificity about how and what federal statistics should be archived.[9] The *Fundamental Principles of Official Statistics* were adopted in 1992 by the United Nations Economic Commission for Europe and subsequently were endorsed as a global standard by the United Nations Statistical Commission. The first three of these principles lay out some of the criteria of transparency, access, and accountability (italic emphasis added):

> Principle 1. Official statistics provide an indispensable element in the information system of a democratic society, serving the Government, the economy and the public with data about the economic, demographic, social and environmental situation. To this end, official statistics that meet the test of *practical utility* are to be compiled and made available on an impartial basis by official statistical agencies to honour citizens' entitlement to public information.
> Principle 2. To retain *trust* in official statistics, the statistical agencies need to decide according to strictly professional considerations, including scientific principles and professional ethics, on the methods and procedures for the collection, processing, storage and presentation of statistical data.
> Principle 3. To facilitate a correct interpretation of the data, the statistical agencies are to present information according to scientific standards on the *sources, methods and procedures* of the statistics. (p. 2)[10]

The FAIR (Findable, Accessible, Interoperable, and Reusable) principles provide further guidance. The FAIR principles focus on making data products accessible and useful, but do not specifically address questions of preservation or providing the metadata necessary to evaluate data quality

---

[8] https://www.whitehouse.gov/wp-content/uploads/2019/06/M-19-21.pdf.

[9] According to the OECD, "Official statistics are statistics disseminated by the national statistical system, excepting those that are explicitly stated not to be official." See https://stats.oecd.org/glossary/search.asp.

[10] https://unstats.un.org/unsd/dnss/gp/FP-Rev2013-E.pdf.

(as opposed to, for example, the metadata necessary for data to be reusable by a third party or to be interoperable with other data resources). Data accessibility is fundamental to public confidence, including the public's willingness to invest in federal statistics. For the transparency of federal statistics, the FAIR principles are a starting place.[11]

ARCHIVING HISTORY AND PRACTICES

A high-quality archive (or archives) is one that meets standards for best practices for preservation, discoverability, and accessibility—of all its data products, including microdata, aggregates, and indicators.[12] Thousands of official depositories at libraries around the country previously provided such archiving for public-use federal statistics, that is, aggregates and indicators. These official depositories made it possible for people across the country to reconstruct what other people believed or knew at a particular time. They also meant that statistical agencies did not consider comprehensive and consistent archiving as their responsibility. Those government document libraries largely no longer exist or are not maintained.[13]

In today's virtual world, it should be relatively easy and inexpensive to preserve data and make them accessible. Appropriate and sustainable institutions are required to do what the official repositories previously did. Because digital data can be altered in ways that are not transparent, it is now often not possible to find out precisely what was published at a particular point in time. More generally, there is no systematic preservation of the public-use digital data products of the federal statistical system. What is preserved is ad hoc and decentralized, and it is often not discoverable by the public. So, while FAIR may be a minimal standard for the transparency of federal statistics, it is not yet being met.

Records (control) schedules (see below) obligate federal agencies, including statistical agencies, to make decisions about the preservation of certain data products (44 U.S.C. 3101 and 3103). The schedules are reviewed in conjunction with archivists at NARA, a "disposition authority number" is assigned, and disposal (destruction) or preservation then follows that

---

[11] See https://www.go-fair.org/fair-principles/ and http://www.nature.com/articles/sdata201618. See also the White House Office of Science and Technology Policy memorandum, Increasing Access to the Results of Federally Funded Scientific Research, http://www.whitehouse.gov/sites/default/files/microsites/ostp/ostp_public_access_memo_2013.pdf, which calls for preservation and access to federally funded research data products, to increase their transparency as well as the efficiency of the federal research enterprise.

[12] Best practice standards for archiving are captured by the Core Trust Seal: https://www.coretrustseal.org/.

[13] See for instance the work put into collecting data for the Bracero program, https://doi.org/10.7910/DVN/DJHVHB.

schedule. Historically, records schedules were poorly monitored, and many of the preservation decisions still appear to be focused on final output products, not necessarily on materials that are of value to research and future activities of statistical agencies, such as metadata and paradata. However, recently there has been an effort to better organize and review records schedules, including searchable catalogs of records schedules and the preserved assets and databases at NARA[14] (see also the discussion below on findability), and the incorporation of records schedule principles into the regular training of federal employees at all levels. Because more data can be more easily captured today, a more expansive view of what constitutes a "record" (44 U.S.C. 3301) might allow for greater preservation of metadata and paradata, which have been traditionally either ignored or identified as "temporary" data, to be destroyed after only a brief delay.

A more complicated question is whether it is important to archive the underlying raw data and programs that are used to produce published statistics. Many recent examples of research have leveraged deep archives of historical data to obtain new insights. For example, the Census Bureau still has the responses to most of the Decennial Censuses, most but not all of which have been digitized. When made public after 72 years, these data have become the most used Census Bureau products ever, sustaining businesses large and small in the family history (genealogy) industry. Social, economic, and methodological research has also used the underlying raw data, even before they are made public. For example, the Decennial Census Digitization and Linkage Project is using the underlying microdata files from the Decennial Censuses between 1940 and 2010 to create a longitudinal data resource that will provide a statistical infrastructure for studying the U.S. population in heretofore unimagined ways (e.g., see Genadek and Alexander, 2019; Genadek et al., 2018).

Research can leverage such raw data, even when it is not digitized. Robert Fogel and Douglas North won the Nobel Memorial Prize in economics for their work analyzing data from federal statistical agencies that had been archived and that they digitized, transforming our understanding of American history and economic development. Various research teams have used data originally preserved by the National Archives, but subsequently (partially) digitized by nongovernmental parties (Ancestry.com, Church of Latter Day Saints, and the IPUMS [Integrated Public Use Microdata Series] project to look into intergenerational linkages (Abramitzky et al., 2019). Research into reclassification errors—see Box 3-1 for a recent example—and how they might affect official or other statistics using the microdata

---

[14]Record control schedules: https://www.archives.gov/records-mgmt/rcs/schedules/index.html; databases: https://aad.archives.gov/aad/; search for preserved assets: https://aad.archives.gov/aad/.

> **BOX 3-1**
> **Recent Classification Issue at the Bureau of Labor Statistics**
>
> The Bureau of Labor Statistics (BLS), which measures unemployment and more general labor market statistics using the Current Population Survey, had to address measurement issues during the 2020 COVID-19 pandemic. Due to the various shutdown measures, unemployment spiked, but an open question was: *How well was it being captured?* The agency itself noted in its report on June 5, 2020, that "large numbers of workers ... were classified as employed but absent from work," leading to a classification error bias in estimates of the official unemployment rate. When employment increased in the June 5 report, there was suspicion—on both sides of the political spectrum—that the numbers were manipulated.
>
> Economists countered such claims with confidence that the BLS "acted with ... integrity and transparency."[a] Erica Groshen, a former BLS commissioner now at Cornell University, noted, "the agency highlights unusual issues in the data, allowing economists and other observers to adjust the numbers as they see fit."[b] In its "Employment Situation News Release"[c] and associated FAQ,[d] BLS points to its procedures ("As is our usual practice, no ad hoc actions were taken to reassign survey responses; the data were accepted as recorded.")
>
> BLS also provided a hypothetical alternative unemployment rate, the rate that would have resulted if all records had been correctly classified ("the overall unemployment rate would have been about 3 percentage points higher than reported"), and the agency let researchers, as always, access the released anonymized microdata[e] or the confidential data, via the federal statistical research data centers (FSRDCs), and come to alternate conclusions.
>
> NOTES and SOURCES:
> [a]Furman, cited in New York Times, June 8, 2020, available at https://www.nytimes.com/2020/06/08/business/economy/jobs-report-data.html.
> [b]Furman, ibid.
> [c]https://www.bls.gov/news.release/archives/jec_06052020.htm.
> [d]https://www.bls.gov/cps/employment-situation-covid19-faq-may-2020.pdf.
> [e]https://www.census.gov/programs-surveys/cps/data/datasets.html.

have been conducted in the past, using both public microdata (Hirsch and Schumacher, 2004) and confidential (raw) survey data (Larrimore et al., 2008).

The National Archives has Manufacturing Censuses from 1929, 1931, 1933, 1935, and 1937 that have been the basis for valuable research, as were the manufacturing and agricultural schedules collected by the U.S. Census Bureau.[15] Some have been digitized and are available through the

---

[15]Examples include Raff (1988); Bertin et al. (1996); and Benguria, Vickers, and Ziebarth (2020).

Inter-university Consortium for Political and Social Research (ICPSR).[16] More recent U.S. economic censuses and surveys are still held by the Census Bureau and made available to the research community through the FSRDC.[17]

The lack of systematic documentation and preservation of digital data continues to create challenges and increase costs—and result in the loss of valuable data. For example, when the U.S. Census Bureau was retiring VAX machines in the early 2000s, it discovered hundreds of data files from surveys undertaken in the 1960s and 1970s for which there was little or no documentation. With the assistance of IPUMS and ICPSR, and leadership from the Census Bureau, these files were preserved, and handwritten documentation was digitized, but much of these data still remain unusable.

Archiving might mean preserving and making accessible the administrative data, including tax data, which are used to produce the Economic Census. Transparency would require that there be public-use documentation of the provenance of the underlying data. Archiving would mean preserving the survey data, the administrative data, and the processes that were used to integrate the two sources of data to create the public estimates. Even in cases where the underlying microdata could not be made publicly available, the legitimacy of the resulting public estimates would be increased if the process by which they were produced and the way multiple data sources were integrated were publicly available.

Becker (2015) and Becker and Grim (2011) report on efforts (and difficulties) to recover usable microdata from the 1950s, 1960s, and 1970s that had not been preserved in the National Archives.[18] They encountered incomplete nondigital metadata on the digital data, failing last-of-its-kind hardware, and unique file formats—all issues that a proper archive with preservation and curation strategies is designed to address. Note that in addition, identifying the precise strategies used to recover the data increased confidence that the analytical results in the research were relevant and meaningful, a consequence of transparency.

Basker et al. (2019) reports on how the historical findings led to new questions concerning contemporaneous surveys. Four new lines of inquiry were added to the 2017 Economic Census regarding (1) retail health clinics,

---

[16] Examples of ICPSR projects using Census of Manufacturing data from the 1930s include Vickers and Zeibarth (2018), available at https://www.icpsr.umich.edu/web/ICPSR/studies/37114/versions/V1; and Bresnahan and Raff (2018), available at https://doi.org/10.3886/ICPSR37208.v1. An example of earlier digitized Census of Manufacturing (1850–1870) data was done by Atack, Bateman, and Weiss (2006); see https://doi.org/10.3886/ICPSR04048.v1.

[17] U.S. Census Bureau microdata from its economic censuses and surveys that are available for research access are described at https://www.icpsr.umich.edu/web/pages/appfed/index.html.

[18] Additional information is available in the Center for Economic Studies Annual Report for 2009, Ch. 4 (https://www2.census.gov/library/publications/2010/adrm/ces/2009-research-report.pdf).

(2) management practices in health care services, (3) self-service in retail and service industries, and (4) water use in manufacturing and mining industries. These were proposed by economists from the Census Bureau's Center for Economic Studies to fill data gaps in current Census Bureau products concerning the U.S. economy. The new content addresses such issues as the rise in importance of health care and its complexity, the adoption of automation technologies, and the importance of measuring water, a critical input to many manufacturing and mining industries.

Today the Census Bureau and other statistical agencies rely on administrative data that are not obtained through surveys and are covered by different laws (for example, the Family Educational Rights and Privacy Act and Title 26 of the IRS Act) that may exempt data from being archived at NARA and may sometimes even prohibit it. Modern surveys often use digital-only survey forms, which may be both adaptive in their structure and dynamic over time, but with no printed "static" equivalent form. These surveys create novel and sometimes unresolved challenges to an agency's ability to archive and preserve its raw input data, processes, and methods.

CURRENT PRACTICES WITH RECORD SCHEDULES AND DATA MANAGEMENT PLANS

As mentioned previously in this report, and as is well known, currently the great majority of the input datasets for official statistics in the United States are either survey based or administrative records based (often from tax data, which are extremely sensitive). These input data sources contain personally identifiable information, and as such can only be made available to the public in an electronically secure environment, such as a federal statistical research data center. In those cases where the input data must remain confidential, what is archived through use of metadata is a description of the variables that the datasets contain and how those datasets are structured.

Identifying which files to preserve, in what formats, and for how long is critical for overall transparency. Ideally, these decisions are made when data are generated, not after the fact. Such documents—defining the how, why, and for how long—exist in a variety of formats. We describe here two dominant examples: *Data Management Plans (DMPs)*, used primarily for researcher-generated data, and *records schedules*, used by U.S. government agencies. They are similar in scope, albeit not in the detail captured.

Data Management Plans

A DMP is a knowledge management document, prepared initially as a specific research or survey project is being planned, to lay out types of data to be collected, the possible presence of sensitive data, the roles of project members in relation to the data, and the planned archiving and preservation of the data. A DMP is a living document that may change many times over the course of the research or survey project. All changes in a DMP need to be documented and retained.[19] Federal research funders have required such DMPs from certain disciplines for some time, but are increasingly expanding the scope of such requirements.

The *Proposal and Award Policies and Procedures Guide*[20] of the National Science Foundation (NSF) notes a requirement for (two-page) DMPs but only as a supplementary material, leaving the evaluation thereof in the realm of individual divisions and program managers. In 2019, NSF encouraged, but did not mandate, the use of DMPs.[21] Discipline-specific criteria can be more tightly enforced. The Interdisciplinary Earth Data Alliance offers its tool to provide proof of compliance with NSF Data Policies. The NSF Directorate for Biological Sciences[22] notes that it conducts post-award monitoring of compliance through annual progress reports.

In October 2020, the National Institutes of Health (NIH) issued its Final NIH Policy for Data Management and Sharing (effective January 2023), which "promote[s] the management and sharing of scientific data generated from NIH-funded or conducted research."[23] Guidance on creating DMPs is provided by numerous entities,[24] and various online tools exist to assist researchers in crafting DMPs.[25]

Records Schedules

All U.S. government agencies are required to maintain "records schedules." All federal records, including those created or maintained for the government by a contractor, must be covered by a NARA-approved agency disposition authority SF 115, Request for Records Disposition Authority, or the NARA General Records Schedules (36 CFR § 1225.10). General Records Schedules (GRS) are schedules issued by the Archivist of

---

[19]NIH defines a Data Management and Sharing Plan as a "plan describing how scientific data will be managed, preserved, and shared with others (e.g., researchers, institutions, the broader public), as appropriate." (84 FR 60398) https://grants.nih.gov/grants/guide/notice-files/NOT-OD-21-014.html.
[20]https://www.nsf.gov/publications/pub_summ.jsp?ods_key=pappg&WT.z_pims_id=0.
[21]https://www.nsf.gov/pubs/2019/nsf19069/nsf19069.jsp.
[22]https://www.nsf.gov/bio/pubs/BIODMP061511.pdf.
[23]https://grants.nih.gov/grants/guide/notice-files/NOT-OD-21-013.html.
[24]See latest guidance from the National Institutes of Health at https://grants.nih.gov/grants/guide/notice-files/NOT-OD-21-014.html. ICPSR also offers guidance at https://www.icpsr.umich.edu/web/pages/datamanagement/dmp/.
[25]Examples of tools include DMPTool (California Digital Libraries), DMPOnline (Digital Curation Centre—UK), and the Interdisciplinary Earth Data Alliance DMP Tool.

the United States (NARA) that authorize, after specified periods of time, the destruction of temporary records or the transfer to the National Archives of the United States of permanent records that are common to several or all agencies (36 CFR § 1227.10). All agencies must follow the disposition instructions of the GRS, regardless of whether or not they have existing schedules.[26]

Government agencies are subject to various rules about retention, in particular the 2011 Presidential Memorandum, "Managing Government Records," and the 2012 Presidential Memorandum of the same title."[27] Agencies are mostly autonomous in deciding what constitutes "records," although the legal code uses the term to refer to information that is "appropriate for preservation [...] as evidence of the organization, functions, policies, decisions, procedures, operations, or other activities of the United States Government or because of the informational value of data in them" (44 U.S.C. 3301). Records schedules can persist for a long time, and are not frequently updated, though recent executive branch memos and efforts at the National Archives may lead to updates and modernizations across U.S. government agencies.

As an example, the surveys conducted by the National Center for Science and Engineering Statistics (NCSES) are captured in the "Request for Records Disposition Authority" form,[28] which specifies that survey forms and questionnaires are destroyed once the survey is finalized; *edited* microdata are transferred to NARA 2 years after completion of the survey, with no disposition noted. The records schedule stems from 1994; there seem to be no other records for the NSF that cover surveys.[29] Raw microdata or paradata are not mentioned, and are thus not preserved by NARA, although NCSES may keep copies for access by researchers.

As pointed out earlier, modern approaches and technical capabilities suggest that even some basic paradata should be preserved (as discussed later in this chapter) as digital traces of the methods and procedures used to create the final statistics. For instance, survey paradata can capture how long respondents linger over questions in digital-only survey instruments, and clear documentation of the edits made to the raw respondent

---

[26] For additional information on record management, see https://www.archives.gov/about/laws (accessed January 13, 2021) and for records schedules, see https://www.archives.gov/records-mgmt/faqs/rcs.html#one.

[27] https://obamawhitehouse.archives.gov/the-press-office/2011/11/28/presidential-memorandum-managing-government-records.

[28] https://www.archives.gov/files/records-mgmt/rcs/schedules/independent-agencies/rg-0307/n1-307-93-001_sf115.pdf.

[29] https://www.archives.gov/records-mgmt/rcs/schedules/index.html?dir=/independent-agencies/rg-0307; and https://www.archives.gov/records-mgmt/rcs/schedules/independent-agencies/rg-0307/n1-307-93-001_sf115.pdf.

> **BOX 3-2**
> **NCSES and Paradata**
>
> NCSES surveys are conducted (in part) by the U.S. Census Bureau. The Census Bureau has a record control schedule specifically addressing surveys it conducts for other agencies (DAA-0029-2013-0002[a]). According to this schedule, it determines that it will delete all partial and final edited response-level data within 10 years, with no permanent record kept at the Census Bureau. The sponsoring agency is responsible for archiving. The record control schedule specifically identifies for deletion at the 10-year mark metadata and paradata, such as "program code, reports, and other items relating to sample design ... questionnaire images, descriptions of weighting methods, ..., interviewer instructions,..." (or sooner, if "the sponsor determines the records are not needed.") Note that the record control schedule for NCSES (under NSF) makes no mention of these files.
>
> In contrast, the record control schedule for the Survey of Income and Program Participation (DAA-0029-2014-0004[b]) treats some of these files differently. Unedited and edited confidential files are destroyed after 25 years, or "when no longer needed, whichever is later." Metadata in the form of "SPIDER specifications for the SIPP Instruments... in XML format" are determined to be permanent (DAA-0029-2014-0004-0005[c]). For the 2005 American Community Survey, even unedited confidential (master) files are kept permanently (N1-029-05-002[d]). It is rare for paradata of any kind to be mentioned in the record control schedules. One example is the post-2007 ACS files (DAA-0029-2015-0001[e]), which mention various paradata such as "Digital recordings of respondents' feedback from scripted probes for each tested item and alternate version of tested items," and specify that these be destroyed after 2 years (DAA-0029-2015-0001-0015[f]).
>
> SOURCES:
> [a]https://www.archives.gov/files/records-mgmt/rcs/schedules/departments/department-of-commerce/rg-0029/daa-0029-2013-0004_sf115.pdf.
> [b]https://www.archives.gov/files/records-mgmt/rcs/schedules/departments/department-of-commerce/rg-0029/daa-0029-2014-0004_sf115.pdf.
> [c]https://www.archives.gov/files/records-mgmt/rcs/schedules/departments/department-of-commerce/rg-0029/daa-0029-2014-0005_sf115.pdf.
> [d]https://www.archives.gov/files/records-mgmt/rcs/schedules/departments/department-of-commerce/rg-0029/n1-029-05-002_sf115.pdf.
> [e]https://www.archives.gov/files/records-mgmt/rcs/schedules/departments/department-of-commerce/rg-0029/daa-0029-2015-0001_sf115.pdf.
> [f]https://www.archives.gov/files/records-mgmt/rcs/schedules/departments/department-of-commerce/rg-0029/daa-0029-2014-0005_sf115.pdf.

data enhances credibility, countering suggestions of improper or erroneous data manipulation (see also Box 3-2, NCSES and Paradata). NARA keeps only those federal records that are judged to have continuing value; these amount to about 2 to 5 percent of the records generated in any given year.[30]

---

[30]https://www.archives.gov/publications/general-info-leaflets/1-about-archives.html.

THE ROLE OF CATALOGS AND SEARCHABLE METADATA

Preserving data and related records does not in and of itself provide transparency if these materials are not themselves discoverable. There are multiple ways national statistics become discoverable, including through the National Archives and data.gov.[31] Some statistical agencies have created their own libraries or archives, such as the National Transportation Library. Specialized archives can be preferable if they are able to make more transparent the embedded knowledge about the data held by particular agencies. On the other hand, given the decentralized nature and limited resources of the U.S. statistical agencies, creation of multiple archives can also *decrease* discoverability, increase costs, and discourage adoption of best practices. In each agency, it will be important to balance the benefits from leveraging their local knowledge and their ability to reach their user communities with the benefits from coordination and integration and the leveraging of existing archival resources. One way to maintain the benefits of decentralization while promoting discoverability is to encourage the use of common metadata standards that facilitate federated searching (e.g., the use of the Data Catalog Vocabulary [DCAT][32] and Schema.org standards makes data discoverable through Google Dataset Search). For details, see Chapter 5.

An inventory of all data resources for each cabinet or other federal agency has been required by policy for nearly a decade, starting with the 2013 OMB memo M-13-13 and now codified by the Foundations for Evidence Based Policy Act of 2019 (now 44 U.S.C. § 3511; hereafter referred to as the Evidence Act of 2019). The M-13-13 policy established the federated architecture whereby each agency compiles an enterprise-wide inventory of all its data assets, using a common metadata schema published as a single JSON file on its website, so that Data.gov and other aggregators can assemble and update combined data catalogs in an automated way. The metadata schema required by M-13-13 was originally referred to as the Project Open Data Metadata Schema, but was revised in 2014 to align with the W3C (World Wide Web Consortium) DCAT metadata standard and is now known as the DCAT-US schema to denote that it is compatible with the international DCAT standard used by many other catalogs.

The DCAT metadata standard is used by a number of other national governments, including those countries following the EU-managed DCAT-AP standard, and it serves as the basis for the Schema.org Dataset schema used by Google Dataset Search and others. Since 2013, most major federal agencies have implemented comprehensive dataset inventories following the metadata standard using metadata management platforms provided by their

---

[31] Data.Gov, while not an archive, does contribute to the harmonization and preservation of metadata and improves the discovery of federal data resources.

[32] For a description of DCAT, see https://www.w3.org/TR/vocab-dcat-3/.

agencies or by Data.gov. To support the federal policy and other DCAT-based requirements, many commercial and open-source metadata management tools have been updated to support the DCAT-based metadata standard. The Evidence Act of 2019, as shown in Box 3-3, codified the federated model from the 2013 policy in statute and expanded the scope beyond the 24 CFO-Act agencies to also include other federal agencies. This data inventory policy and process is one of many in the federal government. For example, the 2018 National Geospatial Data Act also codified in statute another long-established OMB policy, Circular A-16, regarding geospatial data and metadata. Nevertheless, the Evidence Act of 2019 is noteworthy for how comprehensive and widely implemented it is.

The basic DCAT-based metadata captured in the data inventories following M-13-13 and the current Evidence Act requirements are unlikely to capture the rich detail needed for adequate transparency for statistical products. However, since these metadata records are required for all data products, and since most agencies have implemented the metadata management systems and processes to support them, they present a useful starting point. They will enable metadata managers to add additional metadata by reference and for researchers to find the core metadata records in the first place. Additionally, metadata standards like DCAT are widely used and meant to be extensible, with domain-specific variants like GeoDCAT and StatDCAT being explored by other national governments. The widespread use of DCAT and its Schema.org variant across the Internet also helps compatible federal metadata records become discoverable through larger aggregators like Google Dataset Search.

An inventory of all data resources for each cabinet or other federal agency is required by the Evidence Act of 2019, as shown in Box 3-3. (Chapter 5 discusses the use of metadata which would make inventories of such resources easier to use.)

Recommendation 3.1: The agencies that produce federal statistics, through the leadership of the Interagency Council on Statistical Policy and the Chief Statistician of the United States, should fully comply with federal record schedules, ensuring that the input datasets that can legally be retained, and official estimates that are produced, are archived in the National Archives and Records Administration. The metadata that accompany such data should also be preserved using broadly accepted metadata standards appropriate to the data at hand. The records schedules, which describe the plans for retaining, preserving, and making accessible microdata and associated metadata, should be easily accessible on each statistical agency Website so that users know when and where microdata and associated metadata will be made available, and when they are scheduled to be destroyed.

**BOX 3-3
Excerpts from 44 U.S.C. § 3511:
Data inventory and Federal Data Catalogue
(emphasis added)**

(a) Comprehensive Data Inventory.—

(1) In general.—
In consultation with the Director and in accordance with the guidance established under paragraph (2), *the head of each agency shall, to the maximum extent practicable, develop and maintain a comprehensive data inventory that accounts for all data assets created by, collected by, under the control or direction of, or maintained by the agency. The head of each agency shall ensure that such inventory provides a clear and comprehensive understanding of the data assets in the possession of the agency.*

(2) Guidance.—The Director shall establish guidance for agencies to develop and maintain comprehensive data inventories under paragraph (1). Such guidance shall include the following:

(A) *A requirement for the head of an agency to include in the comprehensive data inventory metadata on each data asset of the agency,* including, to the maximum extent practicable, the following:

(i) A description of the data asset, including all variable names and definitions.
(ii) The name or title of the data asset.
(iii) An indication of whether or not the agency—
(I) has determined or can determine if the data asset is—
(aa) an open Government data asset;
(bb) subject to disclosure or partial disclosure or exempt from disclosure under section 552 of title 5;
(cc) a public data asset eligible for disclosure under subsection (b); or
(dd) a data asset not subject to open format or open license requirements due to existing limitations or restrictions on government distribution of the asset; or
(II) as of the date of such indication, has not made such determination.
. . .
(v) *A description of the method by which the public may access or request access to the data asset.*
(vi) *The date on which the data asset was most recently updated.*
(vii) *Each agency responsible for maintaining the data asset.*
. . .

(x) The location of the data asset.
(xi) Any other metadata necessary to make the comprehensive data inventory useful to the agency and the public, or otherwise determined useful by the Director.

. . .

(D) A requirement for the head of each agency, in accordance with a procedure established by the Director, to submit for inclusion in the Federal data catalogue maintained under subsection (c) the comprehensive data inventory developed pursuant to subparagraph (C), including any real-time updates to such inventory, and data assets made available in accordance with subparagraph (E) or any electronic hyperlink providing access to such data assets...

(c) Federal Data Catalogue.—

(1) In general.—
The Administrator of General Services shall maintain a single public interface online as a point of entry dedicated to sharing agency data assets with the public, which shall be known as the "Federal data catalogue". The Administrator and the Director shall ensure that agencies can submit public data assets, or links to public data assets, for publication and public availability on the interface.
(2) Repository.—*The Director shall collaborate with the Office of Government Information Services and the Administrator of General Services to develop and maintain an online repository of tools, best practices, and schema standards to facilitate the adoption of open data practices across the Federal Government*, which shall—
  (A) include any definitions, regulations, policies, checklists, and case studies related to open data policy;
  (B) facilitate collaboration and the adoption of best practices across the Federal Government relating to the adoption of open data practices; and
  (C) be made available on the Federal data catalogue maintained under paragraph (1).

. . .

(d) Delegation.—

The Director shall delegate to the Administrator of the Office of Information and Regulatory Affairs and the Administrator of the Office of Electronic Government the authority to jointly issue guidance required under this section.

SOURCE: 44 U.S.C. § 3511.

ISSUES ARISING WITH PARADATA

While our focus has been on the need for transparency of metadata in the survey collection process, we note that data *about* the data collection process, so-called "paradata," can be important for understanding the quality of the data collected as a function of elements of that process. Although the paradata concept was branded almost 25 years ago (see, e.g., Couper, 1998), the analysis of such data and the appropriate inferences one can draw from doing so continue to evolve. Kreuter (2013) describes paradata as "additional data that can be captured during the process of producing a statistic." Such data are obtained throughout the survey process—as part of the initial interaction, the field staff's observations, and the respondent's actions. The data can also be used to help ascertain and improve the quality of the collected data. In addition, it must be stressed that these data may be sensitive, because they are tied directly to an individual respondent and therefore their release can lead to a breach in confidentiality.

Currently, there is no clear dividing line that separates paradata and metadata. But it is fair to assert that some metadata arise from or are aggregate paradata (e.g., response rates are a summary of individual data on participation or refusal decisions at each call).

The cases below illustrate the nature and level of detail of paradata. We include a short discussion of them here, because these data are collected as part of the production process and can inform survey data improvements and assessments of survey data quality, most often internal to a statistical agency. As a result, maintaining and understanding these data can be viewed as a component of an agency's transparency about the information it makes available.

As described above, in addition to knowing which mode of data collection was used, what measurement instruments generated the data, and (or) detailed information about the population (i.e., metadata), having information about the actual data-generating process—the paradata—will increase transparency. In situations where the data-generating process is in the hands of statistical agencies or other well-defined entities and is designed for a specific purpose (i.e., a survey), systems can be put into place to capture sufficient information about that process itself. Even if the design is not under the control of the agency, effort should be made to capture and retain relevant information to evaluate the data generating process (while the data collection is happening or post hoc). (We note that the methods of Chapter 5 address the means for organizing paradata for archival purposes by conforming to another metadata specification.)

Consider the example of a survey-data production process. It is a process that involves many actors making impromptu decisions, informed by observations from the ongoing data collection process and affecting the final product. For example, when addresses are canvassed, listers walk or drive around neighborhoods and decide about the inclusion or exclusion of certain housing units. Those managing field work use their experience and judgment to decide which interviewers to send to which locations, where to intensify efforts, and where to reduce them. In face-to-face interviews, ample discretion is given to interviewers in most surveys as to how often to approach a household and how and when to engage with a specific housing unit. During the interview, respondents take varying amounts of time to answer a particular question, additional people might be in the room observing and influencing the interview, or the interviewer might probe in specific ways that affect respondents' answers.

Getting feedback from all of these actors about the data-generating process would be a tall order. However, given that most survey-data production processes are aided by digital devices, over the last decade paradata have been collected and used not only to guide process decisions but also to evaluate the quality of the data afterwards.

Case Studies

In this last section we review three types of paradata whose capture and preservation may prove highly useful: data concerning interviewers, call record data, and data derived from measurements of respondent survey behavior such as keystrokes.

Interviewer IDs

Many products produced by statistical agencies still rest on interviewer-administered surveys. In these surveys, interviewers influence greatly the collection and processing of the collected information, so capturing paradata about the interviewers and their activities is necessary to ensure transparency. A recent publication on the effects interviewers have on the statistics product (Olson et al., 2020) demonstrated the variety of influences interviewers can have not only in interviewer-administered surveys but also in a variety of mixed-mode surveys. These surveys may rely on interviewers, for example, to list housing units, to create or augment a sampling frame, to contact and persuade sampled cases, to record additional information about the sampled units (including, for example, biomarkers, specimens, and purchase receipts), or to code answers following the survey. Consequently, even if interviewers are not involved in the actual administration of a survey, there is ample opportunity for correlated response variance due to specific interviewer behaviors (Hansen, Hurwitz, and Bershad, 1961; Kish, 1962), or due to biased responses (Schaeffer et al., 2018).

Measurement error corrections, and investigations into quality, are only possible if an anonymized interviewer ID becomes a standard survey release item. The survey methodology community refers to such information as paradata, distinguishing these types of information from the actual survey content and metadata, which typically capture more aggregated descriptions of the data collection process (i.e., response rates), rather than information at each row of a dataset.

Call Record Data

Many survey data collection operations within the federal government use contact history instruments or case management systems designed to capture information about what happened during the attempts to contact a sampled case, including the length of an interview. Interviewers can record whether a successful contact was made, who responded, and what reasons were given for nonresponse, or make additional observations about the household (especially during face-to-face contact attempts). Methodological research has demonstrated such data to be of sufficient quality to guide the data collection process or to investigate failures afterwards (Bates et al., 2010; Lavrakas, 2008; West et al., 2020).

Examining contact attempts jointly with interview length and interviewer ID information has allowed several survey products to uncover fraud or other problems with specific cases or interviewers (Schwanhäuser et al., 2020). Some agencies have started to make use of call records to convert to more responsive data collection designs that are more cost efficient (Chun, Heeringa, and Schouten, 2018). However, many agencies lack the internal resources to do so, and access to call record data (as well as other paradata) would allow external researchers[33] not only to evaluate the quality of the final survey product, but also to make suggestions on how the data collection process could be improved. Currently it is an open question how large the user base for paradata from call record data is, and this needs to be weighed against the effort to make such data available. A good example of disseminating nonresponse-related paradata is the National Health Interview Survey (Dahlhamer and Simile, 2009; Maitland, Casas-Cordero, and Kreuter, 2009).

---

[33] Adam Eck, then a graduate student, had an internship at Census with the goal of obtaining those data for external analyses. As a result, he wrote a program to deidentify those data internally at Census that passed the disclosure review board, and access to a large number of audit trails was made available. This program, presumably, could be used to release audit trails more generally from the ATUS, but we are unaware of its application to date.

Keystrokes and Other Measurement-Related Paradata

Keystrokes and other measurement-related paradata, generated as a byproduct of computer-assisted data collection, were the first processes referred to as paradata by Mick Couper in a presentation at the Joint Statistical Meeting in Dallas (Couper, 1998). Respondents or interviewers leave electronic traces as they answer survey questions, captured through keystrokes and mouse clicks. Today, when computer-aided data collection is even more common, information from respondent's actions, such as keystrokes (outside of time stamps), mouse movements, or voice recordings, can easily be captured. Nevertheless, the systematic analytic use of such information is still less common (Callegaro, 2013; Kunz and Hadler, 2020; Olson, 2013). This could be because their use is not as well established, or because their storage and dissemination are even more difficult. In the American Community Survey, for example, while one unit of observation can represent upwards of 20 contact attempts, a full interview likely has hundreds of keystrokes submitted whether in a computer-aided interview setting or by the respondents in self-administered Web settings. Few examples exist where log data or audit trails are available as public-use data. A notable exception is the American Time Use Survey (Belli et al., 2019), where a dataset has been created containing the interactions by interviewers with a computer-assisted telephone interview (CATI) instrument while entering responses provided by respondents. Its metadata show the expansive size of such files: in a study where 13,200 people were interviewed, paradata describing the interactions with each question amounted to 2,061,889 records. Also, for the National Assessment of Educational Progress's Mathematics Assessment, keystroke and timing paradata from approximately 29,000 students resulted in nearly 16 million records.

To the research community, it is currently unclear for which of the surveys such paradata are available and what the possibilities are if researchers are given access. Since there is a finite set of operating systems and data collection products for CATIs, CAPIs (computer-assisted telephone and personal interviews, respectively), and Web surveys, it is conceivable to create standardized code to extract relevant indicators from the broad set of paradata to store alongside the responses. Retaining all relevant individual raw data indefinitely may be feasible.

One possibility is to develop guidelines for the retention of some paradata. As an example, a small, common denominator of paradata availability guidelines that seems feasible is outlined in Box 3-4.

The case studies above illustrate uses of paradata and suggest that their availability, with appropriate documentation, can help survey researchers understand the survey-data collection process. This can be especially valuable for continuing ongoing data programs where research on a data

> **BOX 3-4**
> **Examples of Guidelines for the Retention of Paradata**
>
> A gross sample file with the following indicators:
> - Response/nonresponse
> - Total number of contact attempts to a sampled unit
> - A flag for "ever refused"
> - Mode of data collection for a mixed-mode study
> - Some basic domain indicators (regions of the country, for instance)
> - Paradata expressly captured during a study to evaluate treatment variation for cases.
>
> For responding cases:
> - Edit and imputation flags where appropriate
> - Total duration of the survey administration in minutes
> - If interviewer administered, an anonymized interviewer ID
> - Any interviewer observations or notes collected as well.

collection cycle can help improve the data collected in a future data cycle. The use of paradata, mostly motivated by survey methodologists, has grown over paradata's brief history. Nevertheless, paradata's use is not widespread, and it is hampered by a lack of documentation, privacy concerns, the need for special processing, cost, and a relatively small research community.

Paradata stemming from agency actions (e.g., data on the type and extent of effort expended to collect responses, such as contact history) might be retained with response data for as long as the response data themselves are retained. The same is true for paradata from case management details, such as cases that were moved from one mode to a more expensive alternative mode. Paradata that involve respondent reactions to attempts at data collection, such as keystrokes or mouse movements that suggest backtracking, or response latency, could suggest a lack of clarity in what is being asked of the respondent, which in turn could result in lower-quality responses. Therefore, there might be a good reason to retain these for a fixed period of time for examination. In Chapter 5, we will argue that metadata standards provide the means for organizing data for archival purposes through conforming to a metadata specification, and this could be useful in archiving paradata as well.

Continued research on the use of paradata is important to improve the quality of survey data in ongoing data programs. NCSES should investigate how their data programs use paradata, identify programs that would benefit from the use of paradata, identify what data are valuable to maintain, determine the length of time such data are available to researchers, and ensure that record schedules include the disposition status of such data. While individual programs have different requirements and uses, for the purpose of transparency NCSES management should develop a policy concerning the availability and use of paradata consistent with its mission.

The federal statistical agencies should retain, preserve, and make accessible machine- and human-readable metadata—including survey instruments and the provenance of any administrative data—used in the production of official statistics. In addition, because paradata help to provide a better understanding of the quality of survey data, the federal statistical agencies should retain, preserve, and make accessible both machine- and human-readable paradata necessary for evaluating data quality. The comprehensiveness of the retained, preserved, and accessible paradata should be necessarily greater for official statistics used in high-profile decision making, such as the allocation of federal dollars. This includes, but is not limited to, the decennial census, the American Community Survey, and the principal federal economic indicators as defined in Statistical Policy Directive No. 3.

Recommendation 3.2: Federal statistical programs, whose inputs include survey data, should make available, for as long as the data are believed to be of interest to researchers, associated paradata to help users assess the quality of the survey inputs.

Further, when such paradata are associated with a statistical program that is used to distribute political power or substantial federal funds (such as the Decennial Census) and the paradata are a key measure of the quality of inputs to such a program, statistical agencies should make public such assessments for relatively disaggregated demographic-geographic domains.

4

Assessments of Quality, Methods for Retaining and Reusing Code, and Facilitating Interaction with Users

INTRODUCTION

This chapter is concerned with three stages in the life cycle of official statistics. First, inputs are obtained, using a variety of methods. The inputs themselves, as well as the methods used to acquire them, need to be characterized, their quality assessed, and their fitness for use documented. The documentation must be made available to internal and external users, and it must be comprehensible. Second, these input data are processed, which can include data cleaning, transformation, coding, aggregation, analysis, model building, imputation, prediction, and so forth. Processing the data is primarily carried out through use of software code, but in some cases expert knowledge and manual processing are required. All processing steps, code, and instructions, including the assumptions made, must be documented. Finally, the data that are the output of the collection and transformation are published, along with pertinent metadata. The data must be usable to a wide spectrum of users, ranging from nontechnical users browsing an agency Website to sophisticated large-scale users wishing to use application programming interfaces to harvest, post-process, and display the data. Each of these segments of the statistical data life cycle is addressed in the three sections of this chapter.

Regarding the quality of inputs to official statistics, it is becoming more common for federal statistical agencies to use alternatives to survey data as inputs to statistics production. The primary alternatives are administrative data, from federal and state programs, and digital trace data from information stored on the Internet and through other technological means of capturing information. This trend is a natural reaction to the increasing rates of nonresponse in sample surveys and hence to their declining quality, and to the increased availability of detailed, timely digital information about nearly every aspect of our lives. But it raises some difficult issues about how one assesses the quality of the resulting estimates, since there is no longer a framework such as total survey error (which one rarely has for all components for survey data either). The quality of the input affects the quality of the resulting official statistics.

In addition, if one is using a statistical model to combine estimates, which is typical for small-domain estimation (e.g., Fay-Herriot models), one needs to know the relative quality of the direct estimates and the model-based estimates one is combining in order to know how much weight each contribution should receive. However, the literature to date on how to assess the quality of administrative data, or digital trace data, is not fully developed, certainly not to the extent that it is for sample survey data. The associated transparency issue, then, is what should be retained when using administrative or digital trace data, to make whatever quality assessments one needs to make in order to permit use of such an approach to estimation. Further, if such quality assessments are carried out by the parent agency, these would need to be shared with the public as part of an effort to be transparent.

The second topic is the permanent retention of software code (and perhaps the software environment) used in data treatment or to implement the estimation methodology. While Chapter 3 discussed archiving of input datasets and official statistics, we have said little about retention of the software code for reuse, for investigation of the quality of previous estimates, or for checking computational reproducibility. It is worth emphasizing that retention of detailed computer code is essential for assessing (computational) reproducibility. There are a number of tools that have been developed that assist in such activities, including tools for collaborative software development, tools for retention of workflow history, and tools for providing a software environment in which to test code. These and other tools can be used for various purposes to support the retention and later reuse of the code originally used to produce a set of official statistics. While these tools are widely used in industry and academia, they are not yet widely used in the federal statistical system. Therefore, the panel thought that it would be helpful to describe what some of these tools can be used to accomplish.

The third topic addresses transparency in data dissemination and involves the extent to which the federal statistical agencies interact with users to find out what they would like to know about the production of a set of official statistics so that they may be best used. This includes what they would like to know about the data collection processes, the data treatment, the estimation processes, and the validation carried out on the official statistics. Users may have strong opinions about the specific data products offered, including access to special tabulations, whether and where time series of the official statistics are available, and whether input datasets are available for study in federal statistical research data centers.

ASSESSING THE QUALITY OF INPUTS USED TO PRODUCE OFFICIAL STATISTICS

Assessing the quality of survey-based input data is a well-understood matter. The idea of total survey error,[1] in place for many decades, lays out a framework for comparing different survey approaches to producing a set of official statistics by assessing the magnitude of the error coming from various sources. The various sources of survey error are initially divided into sampling error and nonsampling error, with sampling error summarized by the variance due to the sample design, and nonsampling error divided into coverage error, error due to nonresponse (both unit and item), measurement error, and processing error. These various sources of error contribute to the variance and bias of the resulting estimates.

The term "fitness for use" is often mentioned when making such comparisons. This is a somewhat more general notion, because it can include additional considerations, such as relevance and timeliness.[2] However, even in this somewhat more general approach, estimating the biases and variances of the inputs going into the production of a set of official statistics is key.

In recent years, due to the higher costs of collecting survey data, primarily as a result of the increasing rates of unit nonresponse, other sources of data are increasingly being used in the production of official statistics.[3] In particular, national statistical offices have increasingly used administrative data to produce official statistics. Administrative data are collected as a byproduct of the administration of a governmental program, often by collecting information to determine eligibility for the program, the size of the benefit, and information to help distribute the associated benefits. In addition to administrative data, also under consideration for use by statistical agencies are data collected from the Internet and other "technology" sources, including transaction data, social media entries, and sensor data, which is referred to here collectively as *digital trace data*. What is anticipated is that federal statistical agencies considering such data sources will first estimate the quality of the resulting official statistics using either survey data, nonsurvey data of the types described here, or (better yet) a

---

[1] See Groves and Lyberg (2010) for an authoritative description.
[2] See Brackstone (1999) for an excellent list of the factors that should be considered.
[3] For an example, see Citro (2014).

combination of multiple sources exploiting the strengths of each individual source, with decisions of which approach to ultimately use based on cost-benefit considerations, taking all the factors of fitness for use into account. For these reasons—as indicated in Tables 7-2, 7-3, and 7-4—being transparent includes providing any information about the quality of the inputs used to produce a set of official statistics, since such information is at times not retained and would inform as to why the approach taken was decided upon, as well as the resulting impact on the quality of the official estimates.

For such an assessment to work, the agencies need to link the quality of the resulting official statistics using these various approaches to the information on the quality of the data from each individual source used. Consider the case of a statistical agency interested in changing from using a survey-based input dataset to one based on administrative data, for official statistics that are estimated means or totals. Such an agency should have to make the argument that some analog of total survey error for the proposed input data had either the same or lower levels of total error as the previous input data, or that the proposed method met other pressing needs (e.g., faster updates or much lower costs) despite an increase in total error. The need for transparency would dictate that all the information contributing to this comparison be made public.[4]

Because a statistical agency using administrative data for input into a set of official statistics will wish to assess the quality of the information collected by the administrative agency for this (often unintended) purpose, and because statistical agencies will not typically have access to any information about the errors that occurred in the collection of such data, it is important for the administrative agency to share any information they may have on the quality of their data. One concern is that the administrative agencies may not have the resources or the staff expertise to undertake an evaluation that provides the necessary information for this alternative use. Another concern is that the administrative agency needs to let the statistical agency know of any changes to their programs that could affect how they are used and their quality. To address this, interagency agreements may include additional data constructs (some type of paradata) that the administrative agency could record and share with the statistical agency.

As mentioned, in addition to administrative data, federal statistical agencies are also considering—and currently use to a modest extent—digital trace data in the production of official statistics. One example is that the Australian Bureau of Statistics is using supermarket scanner data (since 2011) and Web-scraped data (since 2016) in its estimation of that country's

---

[4]See Rancourt (2018), which explores what could be the framework for using administrative data more aggressively, and the embryo of a framework to measure the quality of the resulting estimates.

consumer price index.⁵ As with administrative data, the use of digital trace data in support of a set of official statistics makes it important to evaluate their quality and the result of this on the quality of the resulting official statistics. This requires knowing details about the underlying data-generating process and the resulting fitness for use of the input data.

If official statistics are produced as a combination or integration of several sources of information, how these various sources are best combined will likely be a function of the quality of the various inputs.⁶ An example is the linear combination of direct (often survey) and indirect (model-based) estimates used in a small-area model, such as the empirical Bayes Small Area Income and Poverty Estimates models for states and counties used by the Census Bureau. In such a methodology with survey-based inputs, the proper weight given to the direct and indirect estimates depends on assessments of the relative sizes of their survey error.

In a sense, this kind of comparison is the analysis that Wolter and Hogan (1988) carried out to show that the error in unadjusted census counts was larger than the error in the adjusted counts (though in that case both approaches were survey based).⁷ The problem in doing this is that given the different origins of administrative data and digital trace data, it is not clear what might be meant by an analog to total survey error.⁸ In addition to the lack of clarity as to what the component parts are that need to be measured, or how to measure them, it may not be known whether or how such information should be combined.

How might one proceed? One can assess the quality of an input data set either externally, by comparing performance to that of some "gold standard,"⁹,¹⁰ or internally, by carefully assessing each step of the process through which the input data were collected. If there were a gold standard for the input data, one could sum up the differences and use that as a metric for quality assessment. Unfortunately, there rarely are gold standard comparison values. One is therefore limited to internal assessments,

---

⁵For more detail, see https://www.abs.gov.au/articles/web-scraping-australian-cpi.

⁶For further discussion of how various sources of information are combined, see Lohr and Raghunathan (2017), and Beaumont (2020).

⁷See also Mulry and Spencer (1991).

⁸For a more exhaustive review of data quality domains and dimensions, see Federal Committee on Statistical Methodology (2020).

⁹Depending on what is viewed as a gold standard, one might have to account for differences in concepts, various kinds of mistakes, omissions, etc.

¹⁰In fact, when we conduct a social inquiry (a survey) it is a controlled scientific experiment in which statisticians control the selection method so that variations and levels are caused by the socio-economic phenomenon. When we use administrative data, we lose this control and so the result is that we cannot ascertain with the same conviction that the changes and levels are due to the socio-economic signals; they could be due instead to the selection/participation process.

accumulating what is known about the magnitudes of the various sources of error to which these estimates are subject. Further, it is not clear how these various magnitudes of error are to be combined into a single metric for comparing various types of estimates.

Administrative Data: Estimating Standard Error

It has now become standard practice for federal statistical agencies to use estimates of the standard error of survey-based official statistics, often in the form of coefficients of variation, for standard aggregates, for example, weighted means and sums. Given that considerable variability is often attributable to nonresponse (both unit and item), it is generally considered best practice to include in such estimated standard errors the contributions due to unit and item nonresponse. This is usually accomplished under the assumption that nonresponse is missing at random.[11] The remaining sources of error identified in the total survey error approach are typically not estimated given the difficulty in measuring them.

How should one estimate the standard error of a weighted mean coming from administrative data? First, one can argue there is no sampling error, because one has a census of the population. However, the target population for a program can differ markedly from the population of those receiving the benefits from some governmental program, possibly because not all the eligible individuals are signed up, or some noneligible individuals have been mistakenly included in the administrative file, or some of those in the file may be duplicates, possibly as a result of living at multiple addresses. The result is likely some under- and some overcoverage. There also can be adjustments and imputations in the case of nonresponse. Further, there might be transformations needed to account for differences between what would be collected in a survey and what is collected through administrative data—such as whether there is agreement for the time window of interest and who or what precisely are the members of the target population—and it may not be obvious how to make the necessary transformations from the administrative world to the survey world. For these reasons, it is not a safe assumption that administrative data from a closely related federal program provide information representing the populations of interest to support production of a set of official statistics (see Cunnyngham, 2020).

And how should one estimate the standard error of a weighted mean drawn from digital trace data? It is typically the case with such data sources that some population members have no chance of being represented while others are represented multiple times, with no clear way to determine how many are omitted or overcounted. There are also questions regarding

---

[11] For a review of methods to measure total survey error, see Biemer (2010).

what the intended measurements on the target population are and how the digital trace data correspond to these intents. If a statistical agency was responsible for the data, and the response/participation mechanism was not understood, there would likely be nonresponse studies to find out if the nonrespondents have similar characteristics to the respondents. However, this does not often happen for administrative or digital trace data.

As is obvious, when the official statistics are model-based estimates formed using survey data—say, some type of regression estimates—the errors in the dependent variable and in the predictors can affect the errors in the estimated regression coefficients and therefore the fitted values. In the case of small-area estimation, the standard errors of the domain estimates are estimable if the standard errors in the inputs are (see Prasad and Rao, 1990). However, it is now becoming common, as in the case of the Small Area Income and Poverty Estimates program, for some of the predictors used by federal statistical agencies in their models to come from administrative data. In such cases, given the lack of information on the quality of these data—or, for that matter, the quality of digital trace data or other sources of nonsurvey data—estimating the quality of official estimates can be challenging. This is another reason why transparency is important. These are issues and complexities that have not been fully resolved and will require further research.

In the case of administrative data, some of the recent contributions to this research area are detailed by Zhang (2012), Reid and colleagues (2017), and Oberski and colleagues (2017). Zhang (2012) notes "a clear lack of statistical theories for assessing the uncertainty of register-based statistics" and then provides a good first step for doing so. He divides contributions to error into two parts. The first part is the contribution to error from the use of an individual input dataset in producing a set of official statistics. Errors here may be due to measurement deficiencies, or they may be due to the degree to which the input data are or are not representative of data from the target population. The second part, which is divided into these same two pieces, contributes to error in the integration of multiple sources of data and so is concerned with issues such as using post-stratification weights and matching. Reid and colleagues (2017) expand Zhang's approach by including a third part, the uncertainty from estimation. They point out

> Designing statistical outputs that use administrative data creates many new challenges because we have to give up direct control over many processes, including population definitions, collection methods, classifications, and data editing. Each administrative source has its own particular problems that must be understood both for our own design work and to assure the final users of the data that our outputs are fit for purpose. When we use

administrative data instead of a traditional survey, we need new processes, such as data integration, and recoding or adjusting administrative variables, which can introduce new types of errors. (p. 478)

At the present time, there is no uniform agreement on what to provide to summarize the quality of the information from an administrative data source. Amaya, Biemer, and Kinyon (2020) provide more operational guidance in recommending the use of what is referred to as the total error approach. Even if the approach taken is not comprehensive, whatever information contributes to an understanding of total error is valuable. We have mentioned some of the issues above. It is anticipated that in many cases there will only be minimal information on, say, how often respondents made use of some type of assistance in filling out an eligibility form, which does not get one very far but is a start. Even if this information does not directly lend itself to some assessment of total error, having any information available on data quality might contribute to better use of administrative data when used in combination with other data. One interesting question is whether some sort of summary error assessment could be used in estimating the best linear combination for a small-area estimation model where the direct estimate is from administrative data (see also Benzeval et al., 2020).

Web-scraped Data: Quality and Other Issues

If the primary dataset used to develop a set of official statistics is scraped or otherwise collected from the Internet, the analogy regarding total survey error in association with survey data is much less clear even than that for administrative data. The population of transactions on the Internet or the population of those making transactions is difficult to understand because (1) it is sometimes difficult to know how many times a single individual is represented in such a database, and (2) it is likely impossible to know how many people are *not* represented. (On the other hand, there may be no nonresponse or misresponse.) We do believe that such data will be found to be extremely useful as predictors in statistical models used to produce official statistics, but the assessment of the quality of such official estimates may take longer to understand. The current state of research is summarized in Amaya, Biemer, and Kinyon (2020).

Statistics Canada has a set of principles for determining when and how to use Web scraping, which includes a gold standard that could be developed to measure quality aspects of a Web-scraped data source.[12] The National Center for Education Statistics (NCES) has a good set of frame information for all public[13] and most private[14] elementary and secondary schools and

---

[12] See https://www.statcan.gc.ca/eng/ourdata/where/web-scraping.
[13] Common Core of Data: https://nces.ed.gov/ccd.
[14] Private School Survey: https://nces.edu.gov/surveys/pss/.

postsecondary institutions[15] in the country, which facilitates measuring coverage errors in a Web-scraping project about schools and universities. These sources include data that could be compared with the same identifiers on Websites to check agreement rates or correlations with high-quality data. Inferences can then be drawn on the quality of data unique to the Web-crawling collection.

One hurdle that arises in assessing the quality of much digital trace data is the lack of good population frames like those noted from NCES above. If one is estimating regression coefficients for a model-based estimate that are assumed to be constant throughout the population, and one is using digital trace data for one or more predictors, so long as the dependent variable being fit is from high-quality frame data one can assess and understand the error of such an estimate. On the other hand, if one is trying to estimate a population parameter using only digital trace data, for example using some type of ratio estimate, the error properties of the estimate may be more difficult to assess. It seems difficult to avoid having the data contaminated by undercoverage and overcoverage. As another example, some researchers have tried to make use of search engine data as predictors of dependent variables of interest (e.g., Google flu trends). But various dynamics in how often the public makes use of specific search terms can have a negative impact on the predictive strength of such predictors, which often cannot be anticipated. These issues remain topics of ongoing research.

The Federal Committee on Statistical Methodology (FCSM) has been involved in a multiyear project, at the request of the Interagency Council on Statistical Policy (ICSP), to develop an evaluation framework for integrated data that will help to identify documentation needs. Several workshops have taken place and reports issued (see FCSM, 2020). The committee's final report, *A Framework for Data Quality* (FCSM-20-04), provides important information on defining the components of quality for datasets, documenting components of data quality, and identifying threats to data from these various sources.

Czajka and Stange (2018) provide a review of international standards and guidelines in the reporting of quality for integrated data. In their report, they review the literature on reporting standards in support of transparency for administrative data. They are supportive of the work of Zhang (2012), and they review what is referred to here as digital trace data, where very little work has been done. They also cite the United Nations Economic Commission for Europe (UNECE) Quality Task Team (UNECE, 2014) as the most significant work on the topic. Internet transaction data and social media data represent a substantial amount of data of generally unknown quality or provenance. Czajka and Stange (2018) argue that while there are many kinds of data to be mined from the Internet, the reporting framework is not well developed, and they suggest that much more research and discussion is needed. Until common agreement is

---

[15]Integrated Postsecondary Education Data System: https://nces.edu.gov/ipeds.

reached on a reporting framework that helps data users determine a dataset's fitness for use, data producers should provide a discussion of the known statistical strengths and weaknesses for each statistical application, opportunities for undercoverage, degree of nonresponse, and opportunities for misresponse.

TRANSPARENCY IN PROCESSING, SOFTWARE DEVELOPMENT

As a matter of principle, all data collection, data processing, and data transformation must be made as transparent and open as possible. Currently, most data collection, processing, and transformation is conducted using software, although some manually constructed survey instruments, manual entry of paper forms, and expert input to data cleaning may also be used. This section primarily discusses software tools that assist in making the software code that conducts these actions transparent, but will also elaborate on how to make manual steps as transparent as possible.

Transparency of Computer Code

A survey instrument may be programmed in Blaise,[16] hosted on open platforms such as LimeSurvey[17] or on commercial platforms such as Qualtrics,[18] or developed in some other custom software designed specifically for the purpose of a particular survey. Data cleaning and transformation may be conducted in SAS,[19] SQL, Python, and other tools, and model analysis and imputation may be implemented in the above-named software as well as in Stata, R, SPSS, or other programming tools. For any particular portion of the collection and processing, code may range from several hundred lines to hundreds of thousands of lines of code.

What all of these scenarios have in common is that at least some of the code is developed in house or by contractors, specifically for the purpose of the statistical agency. Even though SAS and Qualtrics are commercial software platforms, the particular code being run or the particular specification of a survey instrument is clear and remains under the control of the agency. At any point when data are collected, transformed, or used in a statistical model using such code or specification (which will be referred to as "code" from now on), a snapshot of the code should be preserved. One reason for the importance of this is that since methodology programs are modified relatively often, knowing which version gave rise to a set of estimates is not always clear, and being transparent about the specific version

---

[16] https://blaise.com/blaise/about-blaise.
[17] https://www.limesurvey.org/.
[18] https://www.qualtrics.com/.
[19] https://www.sas.com/en_us/home.html.

used is crucial in assessing reproducibility. The evolution of the code may also be of interest, primarily for auditing purposes, allowing internal and possibly external users to identify when a change was introduced that might have affected data quality or fitness for use.

One of the most fundamental tools is a version control system. Briefly, a version control system allows anyone to modify and save any part of an arbitrarily large software system (or its documentation), without affecting the work of others. Such capability is typically used to create a process (and tools) to enable other experts to review and test the change and, potentially, leads to that change being incorporated into the released version of the system delivered to end users.

This practice and associated tools have been refined over time to make the entire review, integration (among the many changes made by all developers working on the project), and quality control (in the form of regression testing) into a sophisticated *release management* process. Such release management increases transparency, as all decisions—starting from a developer's change to feedback by the reviewers, and the tests that have been run by the quality team—are all fully documented by the associated tools. Further, in case any issues arise, this kind of release management makes it easy to track how and why every change was made. Needless to say, without the tools designed for release management, this would be an impossible task. More specifically, *continuous integration* has been a critical innovation allowing a developer to immediately check if his or her changes might cause malfunctions if integrated into the system. Such checking is performed by *regression testing*, where automated tests are run on the changed system to ensure that the outputs produced are still correct.

While transparency was not the main objective for commercial software development, it was essential for open-source development. In fact, most of the ideas related to the concept of open science appear to have originated in open-source development. Unlike in older commercial software projects, where development teams used to be co-located, open-source developers are often distributed all over the world. So, there is an inherent need to come up with suitable governance strategies and collaboration tools to produce software in this environment. The key principle in such distributed collaboration is to explicitly document and make public all individual and group decisions and be able to incorporate input from outsiders. The mailing lists for developers typically describe all new requirements for software, as well as changes in governance or in release processes. More technical input is collected through issue trackers that allow anyone to report a problem, a suggestion, or code improvement.

The concepts and tools related to a version control system—including code review, continuous integration and testing, and collaboration and issue tracking—are not magic solutions but require discipline and good practices to be used effectively. Importantly, many organizations build upon these fundamental building blocks to devise good practices for each task. For example, release management typically involves a variety of quality gates (e.g., limiting the number of outstanding issues), criteria for feature selection, and defined roles and responsibilities of the different parties in commercial and open-source projects alike.

Modern version control software, and more generally software configuration management software, has increased substantially in usability and availability. While stalwarts such as Concurrent Versions System were developed in the 1980s, more modern systems emerged in the early 2000s (Subversion, now Apache Subversion, was at one point the most popular system). As of 2021, "Git," created in 2005 by Linux creator Linus Torvalds, appears to be the most popular open-source version control software. Many commercial versions also exist, such as Mercurial (created in 2005), ClearCase (1990), and Visual SourceSafe (1995).

The use of such systems within statistical agencies and their contractors, which appears to have increased in the past 10 years, is not documented in a consistent way, as far as the committee could ascertain. As noted by Rob Sienkiewcz (Census Bureau) in a presentation, the Census Bureau's Longitudinal Employer-Household Dynamics (LEHD) has been consistently using (internally available) version control since 2001, and Subversion since 2003, and it uses a formal release policy for all code used to generate published statistics, though such code is not publicly posted. It claims to be able to identify all the code used in the production of the Quarterly Workforce Indicators for each release going back 15 years. The Decennial Census's Disclosure Avoidance System has been using Git internally for all of its development, using software by Github and Gitlab, and has published the code as used for its 2010 demonstration products in full.[20]

As the LEHD case demonstrates, it is possible to use such systems for audit purposes, and as the Decennial Census's Disclosure Avoidance System shows, it can promote transparency and public inspection of novel technologies. The LEHD code also successfully avoids use of any hard-coded but secret parameters, a feature that is incorporated into each code review of released code, making it technically possible to easily publish the code.

Given the focus on data for federal statistical agencies, we focus on tools that have been developed in the area where software and data intersect. For example, contemporary machine learning and statistical techniques require a plethora of software packages that are to be used in a single data workflow involving data collection, cleaning, and transformation, numerical simulation, statistical modeling, data visualization, machine learning, and much more. One example of the use of supervised machine learning at

---

[20] https://github.com/uscensusbureau.

the Bureau of Labor Statistics is to map the words people use to state what their occupation is, or what injuries they have experienced on the job, into standardized categories.

This multitude of tools may be complicated to install and set up on a workstation. Many such applications are, therefore, shipped as preinstalled virtual machines. This saves time and frustration for someone trying to build upon or reuse existing approaches. Furthermore, with data analysis it is important to tie code and corresponding data output or graphics in a single document. An open-source tool mentioned above, Jupyter Notebook, is a Web application that allows users to create and share documents that contain live code, equations, visualizations, and narrative text and support data workflows integrating tasks ranging from data collection to the presentation of results. Such tools are typically used in conjunction with virtual machines and version control systems. This allows reproducibility even of the most complicated data workflows.

It appears that few of the federal statistics organizations use such tools and practices to engage outsiders or even to coordinate the work within or among the agencies. Using these tools would increase the transparency of the decision making in these agencies prior to the release of data products. Notably, software-as-a-service providers, such as Github and Atlassian, provide an online delivery and sophisticated integration of many of the tools noted above as well as training needed for their effective use. This significantly lowers the entry barriers for federal statistics agencies to transition to a more reproducible and transparent infrastructure.[21]

The Decennial Census program staff uses internal and external versioning systems (Github and Gitlab) for documentation of the code used to perform the computations needed in carrying out the census. Further, they make code publicly available as it is used, up to what can be feasibly released, given various confidential and privacy parameters. Many other statistical agencies use Github for similar purposes.

To evaluate the role these tools might play in the future in the federal statistical agencies, there will be a need for greater access to computer science expertise and an examination of best practices in programming and data curation. The current versioning system of choice is Git, which was created in 2005, so these tools are not recent developments; their value to industry and academia is clear. The additional computer science expertise needed can be acquired through new hires or through various consulting arrangements, with different advantages stemming from each approach.

---

[21] The following Web pages contain useful information outlined in this section: https://git-scm.com/book/en/v2/Getting-Started-About-Version-Control; https://en.wikipedia.org/wiki/Issue_tracking_system; https://www.lucidchart.com/blog/release-management-process; https://www.docker.com/; https://github.com; https://bitbucket.org; https://jupyter.org.

The support of the ICSP and more informal interagency cooperation will be valuable, as some agencies (likely the larger ones) get started earlier on investigations into the role such tools can play and the benefits that doing so provides.

> **Recommendation 4.1:** Agencies that produce federal statistics, including the National Center for Science and Engineering Statistics, should review and make a priority of adopting modern information technology tools that assist in collaborative software development and documentation of workflow and methodology.

This is important, because transparency through computational processes is as important as the transparency of other processes, and also because it will make their transparency efforts more efficient and facilitate internal reproducibility and evaluation.

However, it will not be sufficient to simply collect incomprehensible (if functional) code. Code can contain copious amounts of documentation, can be structured to be more easily understandable without documentation in English grammar (using programming style guides), and can be accompanied by high-level and detailed documentation and software-agnostic specifications. All such documents and practices increase both internal and, when published, external transparency. Many agencies spend many person-hours crafting the latter, but little is known about the use of programming style guides.[22]

As the Decennial Census's Disclosure Avoidance System has shown, it is possible to develop new systems, including those pertaining to sensitive disclosure avoidance systems, in a public and transparent manner. Doing so supports and encourages transparency, while allowing for computational reproducibility in some cases.

At the other end of the data life cycle, agencies generally publish survey instruments (questionnaires), and these have been used successfully to allow interpretation of historical data collections. Applying the principles of transparency laid out here suggests it is not only the hard-coded questionnaires that are of use, but also the coding instructions for such questionnaires. It is in this context that standardization of coding (e.g., exporting questionnaire specifications in Data Documentation Initiative format) becomes a requirement for greater reproducibility. While it is unlikely that a single researcher will re-implement an entire federal survey, one frequently sees researchers re-using certain questions (e.g., the Current Population Survey [CPS] demographic module) or re-using entire supplements (e.g., the Contingent Worker Survey supplements to

---

[22]For an example of coding style guides, see https://google.github.io/styleguide/.

the CPS as replicated by the RAND-Princeton Contingent Work Survey; see Katz and Krueger, 2019). While the availability of re-usable questionnaire specifications is not a necessary requirement to be able to replicate a federal survey, it can greatly reduce the development costs.

Logging

In addition to retaining the versions of the code that were used to collect, treat, and transform the data, an actual "transcript" or log of the actions conducted by the code may also be relevant. From the academic literature, it is well known that all features of the computational environment can affect computations.[23] It is thus critical to record not just the code used to collect or process data, but also the environment in which such code was executed, and (subject to some reasonable constraints) to record logs of such execution as well. In some environments, keeping such logs may be legally required to document authorized access. While it is possible to keep extremely detailed computational traces (e.g., capturing low-level hardware modifications is technically possible), it is generally considered sufficient to keep logs of the software as it is executed.

In certain secure environments within statistical agencies, all use of the software generates a logfile for audit purposes, but in the context of transparency and reproducibility, simpler logfiles may be sufficient. Compared to the data being generated, logfiles are generally much sparser and thus easier to archive together with the generated data, acting as a form of metadata, or in some cases, paradata.

Special purpose statistical software (SAS, SPSS, Stata) has such ability built in, though it is not always enabled. General purpose software such as Python or C++ will not, in general, generate logs unless explicitly programmed to do so. All use of software as it is being processed should generate logs, and this should be enforced through coding style guides. Whenever code is run to produce published output, all logs in the processing sequence should also be archived.

Modern literate programming tools (Knuth, 1992) often promise greater transparency by having the (English grammar) text interspersed with computer code. Examples include Jupyter Notebooks, Rmarkdown documents, and Sweave documents (all using R). This is not unique to such newer tools. Coding style guides at LEHD from 2005 required that programmers use processes similar to Sweave to intersperse legible English

---

[23]For an example of how this can affect computations within a given programming language, see for instance Gould (2011). For the importance of accounting for these issues for privacy, see Garfinkle and Leclerc (2020). For an older but still relevant description of the issue, see McCullough and Vinod (1999).

language and mathematical formulas with functional SAS code.[24] Some of the newer systems (such as Jupyter Notebooks) may be subject to unreliable "out of order" processing, leading to irreproducible results—the exact opposite of the intended outcome (Wang et al., 2020).

Transparency of Manual Processes

It was mentioned earlier that not all data collection or transformation is conducted purely by code. Paper forms, on which data are manually entered, may still be collected in certain circumstances. Qualitative responses, such as industry or occupation descriptions, may be manually coded into industry and occupation codes. In specific cases, expert input may be used to correct or fill in unlikely or implausible data.

These scenarios have two features in common. First, the data can be compared before and after the manual intervention. Thus, documentation can at a minimum point to preserved and possibly archived data that can be inspected to see the effect of manual interventions. Second, none of the manual processing is done in a void. Employees are trained, using instructions, manuals, and guidance. These constitute "human programs"—instructions that humans read, interpret, and implement. And any such instructions are almost always stored as electronic documents, which in turn can be versioned, archived, and made public using similar tools as outlined earlier in this chapter.

> **Recommendation 4-2:** To facilitate transparency, agencies that produce federal statistics are encouraged to develop coding style guides, and to make available documentation and specifications for software systems, subject to any security concerns. Where possible, code (for example used for data collection or processing) should be made publicly available, subject to redaction or removal of confidential parameters, and logs of processing sequences should be archived. Manual processing steps should be clearly identified and documented, and any instructions or guidance given to the staff conducting such manual processing should be archived and made as transparently available as possible.

This might entail a limited amount of redaction, but the existence of any redaction for privacy-preserving purposes should be considered an integral part of the documentation.

---

[24]Also see Lenth and Hjøsgaard (2007), and for the later StatRep package, see Arnold and Kuhfeld (2015).

Of Special Relevance to Chapter 5

A statistical *activity* is *what* statistical agencies do to design, collect, transform, estimate, integrate, disseminate, or archive data under a statistical program (e.g., a statistical survey or census). An example might be top-coding the values for some variable collected by a survey. Given this, a statistical *process* is *how* each of these is achieved within each of the subject-matter areas in each of the federal statistical agencies. Statistical software languages (e.g., SAS, SPSS, Stata, or R) might be used to carry out the edits. The source code for such an editing procedure would contain detailed documentation for the top-coding edit procedure.

These activities and processes are part of what constitutes statistical methodology. The methodological design corresponds to an activity, instructing as to what is to be accomplished by some part of a statistical program. These instructions are sometimes known as specifications. How that activity is carried out, in theory, is an algorithm, the logical steps needed to achieve the requirements in the design. And these logical steps are also sometimes known as procedures. The source code, for example, is the process in practice when it is automated. The source code includes the steps required by the algorithm and the constraints of the programming language being used. Source code written in any of dozens of programming languages can each perform the same algorithm, and many algorithms can satisfy one design. For example, the problem of sorting a set of values alphabetically or numerically—a design—can be accomplished by several well-known algorithms. Each one can be implemented using one of many programming languages.

A statistical program is an implementation of a set of methodologies put into practice (from designs to algorithms to source code), and when one documents those methodologies one is documenting the statistical program. But, as will be discussed in Chapter 5, documentation and metadata are the same thing, so the source code for the automated portion of a statistical program is part of the (detailed) metadata for that program. Exactly where the source code fits into the overall metadata framework for a statistical agency depends on several factors, and these are also discussed in Chapter 5. One important point to make here is that metadata specifications should be part of the software design and development process.

FACILITATING USER INTERACTION WITH STATISTICAL AGENCIES

Transparency in data and methods is important to many users and their needs, especially to know whether the official estimates are fit for a particular use. Similarly, the cost-benefit assessments discussed in Chapter 2 are not possible unless agencies know which measurements are important to the public and the impact various errors would have on various applications of the official statistics. Consequently, part of any effort to provide greater transparency should be taking a more dedicated and systematic approach to understanding what users need.

More broadly, in order to understand and meet user needs in data products, documentation, dissemination systems, and archiving, agencies must develop mechanisms to solicit more frequent input from their user community and facilitate ongoing dialogue with them. A number of the federal statistical agencies have given limited effort to understanding what their users need in terms of transparency, accessibility, and usability of data products to enable optimal use of official estimates and associated input datasets. But there have not been many surveys or other systematic efforts to collect information about users' needs or assessments of satisfaction with documentation, Web pages, and dissemination platforms.

While there is a considerable amount of valuable information on the Web pages of the federal statistical agencies, these resources are often not easily found. This is an important consideration if one is to ensure that metadata and underlying computer codes are going to become useful to users. Moreover, input from data users has not usually been sought *before* the introduction or redesign of new products, dissemination systems, and Websites. As a result, data users have found that some of the interfaces developed to facilitate access and use of estimates are not intuitive or user friendly (an example is the second version of the U.S. Census Bureau's American FactFinder) and Web pages are not easy to navigate. For a more relevant example for the National Center for Science and Engineering Statistics (NCSES), the two expert NCSES users mentioned in Chapter 2 said that new users often need substantial mentoring from more seasoned users to be able to find and access the data they need to answer research questions of interest. Given that funds and staff time are limited, agencies can more efficiently and effectively target these resources if user input is obtained both prior to and during the development of new products, documentation, dissemination systems, and Website redesigns.

One of the primary challenges agencies face in addressing these gaps in user interaction is first identifying the members of their user communities. Agencies can overcome this obstacle by establishing ongoing data user groups through a variety of mechanisms. For example, agencies could offer data users the opportunity to sign up to receive announcements or a periodic newsletter through email, or they could establish online communities—similar to the Census Bureau's American Community Survey Online Community—that users could opt to join for free. The advantage of an ongoing data user group is the ability to communicate directly with users at specific times and with specific information or requests. In contrast, notices that are simply posted on an agency's Website will reach some members of their user community, but only if they happen to visit the Website during a particular timeframe.

Once an agency has established an ongoing data user group with a contact mechanism (email, online community, etc.), there are four activities it could undertake to increase and enhance interaction with its data users that would benefit the agency:

1. Statistical agencies could periodically survey their user communities to determine the kinds of problems they are experiencing accessing and using agency statistics, including ways that each agency could make its Web pages easier to navigate, or ways they could provide the estimates or microdata to better facilitate various often-used analyses, including time-series and cross-sectional analyses.
2. Statistical agencies could survey their user communities to solicit specific input *before* changes are made to data collection techniques, estimates, data products, dissemination systems, or Web pages to ensure that data users' needs will still be met after proposed changes are implemented; and they could also involve members of the user community in reviewing and providing feedback as these changes are actually implemented.
3. Statistical agencies could create a mechanism, such as an online community or forum, that enables members of their data user group to communicate directly with each other—posing and answering questions, providing solutions to user-identified problems, and raising issues for community member feedback. Such a mechanism could reduce the user-support burden on agency staff and be particularly useful for isolated analysts who need more guidance in finding, accessing, and using particular agency data series.
4. Statistical agencies could meet regularly with representatives from their user communities to engage in more in-depth dialogue about ongoing issues and potential future improvements to estimates, data products, documentation, dissemination systems, and the structure and navigation of agency Web pages. This is similar to the various federal advisory committees that several of the statistical agencies currently meet with. Is it possible that they could be made more robust to address user needs? Could they meet more frequently? Could they be enlarged?

Other possible activities that should be considered include:

1. systematically analyzing comments and questions posted to identify common themes and needs for improvement;

2. conducting a "use case" analysis of different types of users and evaluating the utility of the Website for each type;
3. giving users a place to share not only advice and comments but code and user-generated datasets; and
4. instituting outreach to encourage new use of the data.

In addition, there might be value in reaching out to data journalism groups like Investigative Reporters and Editors/National Institute of Computer-Assisted Reporting, which hosts active listservs where they share tips for navigating the confusing data interfaces from federal agencies.[25]

Restructuring Web pages can be a complicated task for the agencies. There are a variety of types of users that one wishes to accommodate, and different structures might be preferable to different user types. For example, a long-time NCSES data user might wish to understand what changes have been made to the most recent version of a survey; a journalist might wish to simply download the current value to support her article; a non-NCSES data user might happen onto NCSES data found in data.gov and as a result visit NCSES's Website to search across topics and subtopics to learn more about an issue of interest; or an analyst might want easy access to an internal archived dataset to reproduce a specific statistic to check on computational reproducibility. One must also be cognizant that there will always be users who are comfortable with the current Website and navigation, and they will find that their experience is disrupted with the implementation of a new structure. However this is addressed, accessibility is crucial. Is information that is provided findable and usable by users? Whatever system is employed to give users access, it must be navigable in an intuitive way.[26] A topic related to the above is the availability of various technical and methodology reports, metadata constructs, and other ways of providing the public with access to other information on the data treatments and methodologies used in the production of a set of official statistics. This can include codebooks and research reports that discuss attempts to make improvements to different aspects of the computations carried out. In NCSES, such reports are often available internally in draft form, but either they are never reviewed and therefore are not made publicly available, or they are viewed as being too technical for public release. We understand that the demand for such documents may be limited, but for a small subset of the user community such documents can be extremely important for providing detailed information to researchers on how data treatments were implemented and how estimates were produced, and in

---

[25] https://www.ire.org/resources/listservs/.

[26] We have provided Recommendation 6.7 to NCSES concerning their need to better understand their users' data needs and preferences; it can be found at the end of Chapter 6.

informing researchers about which aspects are currently being examined for improvements. Further, having access to these will allow other users to gain an appreciation for the care that underlies these programs and their statistical products.

Finally, it is obvious from the content of this chapter that many innovations are coming on line now. For example, the Coleridge Initiative has sponsored a Kaggle challenge on rich text analysis that employs algorithms that will substantially improve the ability to see who in the research community has used various datasets and what they have published. This will create a real-time opportunity for agencies to see how their data are being used, which in turn will help them become more responsive. This helps all statistical agencies meet a requirement in the Evidence Act to get feedback from the public on the utility of their data. More such innovations are on the horizon, and they will have an impact on transparency.

5

Metadata and Standards

INTRODUCTION

In order to ensure that an archived dataset can used in the future, sufficient information must be provided detailing what the columns and rows (in a typical rectangular dataset) represent. For an input dataset, this necessitates providing the units of analysis as defined by the rows, and the variables defining the columns, including what the questions and transformations underlying each variable are and what the various responses are and what they mean. If the dataset contains output estimates, again what the rows and columns signify needs to be stated. To assess the fitness for use of estimates, especially if they are to be used in combination with other estimates in some way, the variability of the estimates (assuming they are the result of a sample survey) due both to sampling error and to nonresponse also needs to be provided. All of this information needed to analyze a dataset is called metadata.

In providing these metadata, it is extremely helpful if the information is structured using standard schemas so that users understand what information is being provided in each portion of the metadata file. When that is done, the users do not have to figure out how to interpret each element of the metadata. Standardized metadata allow automation of data analysis, transfer, and aggregation.

METADATA: THE BASICS

The term *metadata* was invented by several researchers independently around 1970 (Bagley, 1969; Sundgren, 1973). In practice, though, the need for metadata has been around for as long as humans have organized information. Examples include library card catalogs and classification schemes. For many years, the extent of metadata management and usage was due to information professionals engaged in cataloging, classifying, and indexing things; but as the use of the Internet and the Web has grown, the need for digital information has grown along with it. Now, metadata, originally defined as *data about data*, has come to mean any descriptive information about some objects of interest, because it is simple to attach digital descriptions to a list of objects online.

We now define metadata as *data being used to describe some object(s)*. Library books, data and datasets, museum artifacts, telephone calls, and survey questionnaires are all examples of objects that can be described. Data describing a telephone call (e.g., calling number, receiving number, length of the call) are metadata, unless they are used to create a network of callers and receivers, in which case the data are not being used to describe. This expanded view means use of the term metadata indicates a *role* played by some data; it is not a fixed property.

Statistical metadata are data (information) used to describe statistical objects. Statistical metadata are best understood as structured information. In the statistical community, efforts to begin the management and use of metadata to describe data, datasets, and the methodology used to create them began in the 1970s. In the 1980s the efforts expanded to include research data libraries, data archives, and national statistical offices; and they expanded even more widely in the 1990s as the online digital revolution exploded. This expansion in use is continuing, but metadata management has not kept pace with the ever-increasing amount of data available.

As noted above, metadata is a role for data, and metadata are often further categorized as being, for example, *descriptive*, *administrative*, or *structural*.[1] These too are roles that metadata can play for a user or administrator.[2] The roles are as follows:

- *Descriptive metadata* are needed to describe a resource for discovery and identification; examples include basic elements such as "title," "keywords," and other ways of explaining what a digital object is about.

---

[1] From An Introduction to Metadata: https://www.getty.edu/publications/intrometadata/setting-the-stage/.

[2] Also, the categories are neither exclusive, exhaustive, nor definitive. Other roles may be identified as desirable, and each statistical object may perform more than one. These categories can help statistical agencies think about how to organize their metadata usefully. One caveat here: there are indeed niche metadata standards designed to play specific roles. For instance, the PREMIS metadata standard is meant to particularly record provenance and administrative metadata (https://www.loc.gov/standards/premis/).

- *Structural metadata* indicate how compound objects are put together; for example, how pages are ordered to form chapters, or how different observed events were combined to create a "set" of data observations.
- *Administrative metadata* provide information to help manage a resource, such as when and how the resource was created, its file type and other technical information, and who can access it. Administrative metadata can include the following subsets:
  - *Preservation metadata,* which contain information needed to archive and preserve a resource;
  - *Rights management metadata,* which deal with intellectual property rights or restrictions on access or use (note: often embedded in preservation metadata); and
  - *Provenance metadata,* which contain information about where the data came from and how they were created.

Metadata are provided in two major forms: as text in documents or as formal attributes corresponding to characteristics of some classes of objects. For instance, most cans of food can be described by the following: food name, manufacturer, list of ingredients, and nutritional information such as amounts of sugar, fat, and protein. A picture of what the food looks like after preparation is often included on the label as an enticement to consumers. For statistics, a microdataset may be described by name, producer, size in bytes, record length, number of records, file location, link to a data dictionary, and so on. An attribute is the combination of a characteristic and its value associated with some object, and it is descriptive of that object—for instance, a can of food or microdataset. Descriptions may be presented as prose in text format, as a list of the attributes, or as some combination of the two.

Metadata provided as text in documents are called *passive*, and they are *human readable* only. *Machine-readable* metadata are in a form that can be easily processed by a computer. Machine-readable metadata used as input or created as output in systems are called *active*. For example, metadata stored as prose in PDF documents are human readable and passive. Metadata stored in formal databases as numbers, codes, or entries from controlled vocabularies are designed to be machine readable; they are active when used to control the execution of some system in a particular way.

In this report, *statistical metadata* (hereafter, metadata) are data used to describe *statistical objects* (see below). Data used in this role are metadata. This means data are considered metadata when they are used in a certain way, specifically as describing some "things." As a means for comparison, in traditional statistical surveys data are collected from persons in households or business establishments. Data collected about an individual could be used to describe that person. In that case, they are metadata. For statistical surveys, however, the data are also used to represent the properties of similar members of a population, so the data then have a statistical role, not a descriptive one.

Usually, metadata are a product of statistical agencies. The agencies produce metadata to describe the statistical objects they produce, for example data, datasets, and questionnaires. However, data users may also construct additional metadata based on their experiences and uses of statistical data. If a user constructs a new dataset based on integrating data produced at several statistical offices, then the new dataset needs its own metadata. For scientific research data, metadata are particularly important in facilitating the machine readability of a dataset (the automatic use of a dataset by software), data sharing and reuse, and transparency and reproducibility. Recent pushes to make scholarly data that are "findable, accessible, interoperable and reusable" (FAIR[3]) emphasize the critical role that metadata play in achieving all of these goals (Wilkinson et al., 2016).

For computational and statistical research, it is particularly important to record the workflows and methods used in an analysis (Stodden, Seiler, and Ma, 2018). This might be expressed as an executable workflow—a piece of code that could be used to precisely recreate a computational output—or in some other kind of provenance metadata that clearly show how a final data product was created from raw data sources. (Chapter 5 describes tools to preserve and share such code.)

Common Metadata Objects

There are many kinds of objects produced by statistical agencies in the course of their work. They can all be described, of course, but the kinds of objects each agency needs to describe may vary; the selection of the objects depends on specific needs. Generally, the statistical community produces objects of the following kinds:

- Concepts (especially their definitions)
- Questionnaires and forms
- Questions
  - Wording
  - Response choices
  - Flows (for example, skip pattern)
- Instruments (implemented questionnaires)
- Variables
- Value domains

---

[3] For more information, see the GO-FAIR Website at https://www.go-fair.org/fair-principles/.

METADATA AND STANDARDS

- Classification systems
- Code lists
- Datasets
- Data flows
- Sampling plans
- Processing algorithms and systems
  - Editing and validation
  - Coding
  - Allocation
- Estimators
- Imputation methods
- Disclosure control methods, and
- Application programming interfaces (APIs) and other data channels or services.

For instance, metadata about a dataset include descriptions of the dataset itself and the underlying data. For a machine-readable dataset, describing the variables is just as important as describing the set overall. Characteristics typically used when describing variables include these:

- The concept a variable represents (say, marital status)
- Value domain (<s, single>, <m, married>, <sp, separated>, <d, divorced>, <w, widowed>)
- Datatype (in the case of marital status, nominal datatype),[4] and
- Universe (say, adults in the United States).

Once a list of characteristics is developed, it may be reused for describing any dataset elsewhere, and this can be generalized to any kind of statistical object. Reuse supports compatibility by showing when items are similar in some way. Both similarity and lack of similarity are supported. In this sense, comparability is an important consequence of reuse.

Reuse of metadata is achieved by identifying similarities among objects. For instance, in our example above, we apply the concept *marital status* to a variable. There might be many such variables, especially if an agency conducts a single survey that produces a similar dataset on a regular ongoing basis or conducts several surveys that each collect a marital status variable. If the same concept applies to each, one can simply apply its description each time it is needed. This is called the *reusability principle*: describe once, use many times.

---

[4]The nominal datatype is one where the values are categorized with no ordering imposed. Marital status is an example. The set of states in the United States is another.

Metadata Encoding

Metadata can be recorded in multiple ways. The simplest way might be a README file—a human-readable document explaining how a dataset was created, by whom, and why.[5] However, in order to make metadata machine actionable and indexable in a database, it must be *encoded* in some sort of structure for a computer. ISO/IEC 11179, from the International Organization for Standardization (ISO), specifies a structure for metadata (a schema) that supports machine readability and inclusion in some sort of registry or catalog, and this standard has been foundational in the development of many metadata standards and registries. In particular, ISO/IEC 11179's notion of a *data element* (or variable) will come up again later in this report, for example in our reviews of the Data Documentation Initiative (DDI) and the Generic Statistical Information Model (GSIM). A data element is an atomic unit of data, with precise semantics. Even the most rudimentary metadata standards will take care to define a schema for describing the data elements in datasets. However, ISO/IEC 11179 does not specify the encoding format for the metadata. This is because there is a need to support multiple kinds of encoding, as different approaches have different strengths and weaknesses. The three kinds most relevant to metadata system developers are XML, JSON, and RDF.

Schemas are important because they provide a way to learn of the attributes describing a given kind of object, often through automated means. Which attributes are used to describe a variable? Look at the schema. But the schema also provides information about which attributes are related to which others, which attributes are required, and what rules the allowable values for each attribute need to follow. A schema provides valuable information that a description of some object cannot. In some ways, a schema is metadata describing other metadata.

Schemas are developed by analyzing what kinds of information are needed to describe a particular kind of object. For instance, a description of a dataset might differ depending on information needs, especially for transparency and reproducibility. For example, in a system that is a catalog for files containing satellite images, what is required might be the date and time the image was received, the subject of the image (what body or feature is depicted), and the name of the satellite and imaging camera. In that instance, little about the underlying data is needed. If the system is a catalog of statistical datasets, however, much more has to be known about the variables in the dataset. The needs of the system determine the attributes a schema has to define. It is often the case that relevant schemas

---

[5]See https://data.research.cornell.edu/content/readme and https://ropensci.github.io/reproducibility-guide/sections/metaData/.

METADATA AND STANDARDS   101

already exist, so the need to develop one independently is greatly reduced. Standards are a good source of such schemas, and schemas already in use by other federal statistical agencies are another.

XML (eXtensible Markup Language)

XML stands for eXtensible Markup Language. XML gained popularity as an encoding format in many libraries, museums, and data repositories in the late 1990s and 2000s, and it persists today. Syntactically similar to the HTML used to encode the display of Web pages, XML was designed to be both machine and human readable. Metadata elements are enclosed within pointy brackets ("<" and ">"), which makes it possible for a computer to easily retrieve all the information recorded within one. Elements can also be embedded within another element, providing a hierarchical structure. Each element has a name associated with it, called a *tag*.

Unlike HTML, XML is user defined. This means the elements and their tags are defined by their needs within the design of some system and are not predefined as they are in HTML. There is no preset list of elements that needs to be used within an XML document; rather, different metadata languages can be written (or referenced) for different purposes. However, element requirements can be encoded in a schema language called XML-Schema, and this feature allows for the same element set to be reused. Figure 5-1 contains an example of a metadata document.

```
<Data Set>
        <DataSetName> Consumer Expenditure Public Use Microdata – Interview </DataSetName>
        <Location> https://www.bls.gov/cex/pumd/data/sas/intrvw20p1.zip </Location>
    <FileType> SAS </FileType>
    <Variable>
            <VariableName> Fam_Size </VariableName>
            <DataType> numeric </DataType>
            <TypeDescription> positive integer </TypeDescripion>
    </Variable>
            <Variable>
                    <VariableName> Fam_Type </VariableName>
                    <DataType> category </DataType>
                    <Category>
                            <Meaning> married couple only </Meaning>
                            <Code> 1 </Code>
                    </Category>
            <Category>
                            <Meaning> married couple, own children only, oldest child
< 6 </Meaning>
                            <Code> 2 </Code>
                    </Category>
            </Variable>
</DataSet>
```

Figure 5-1 Example of a simple dataset description in XML.

In the example displayed in Figure 5-1, the metadata record is describing a public-use microdata set from the Consumer Expenditure Interview Survey from 2020. Note the hierarchical structure of this document in Figure 5-1; every indented node is read as the "child" of the node above. This hierarchical structure makes XML particularly useful for encoding metadata by describing objects that fit a hierarchical, part-whole or parent-child structure, as well as for encoding documents.

JSON (JavaScript Object Notation)

JSON (JavaScript Object Notation) is a data-interchange format. It is lightweight, meaning the syntax for using the language is uncomplicated, easy for humans to read and write, and easy for machines to process. JSON is based on the JavaScript language. It is a text format that is independent of other language but uses conventions that are familiar to programmers. For these reasons, JSON is ideal for data interchange.

JSON has a simple structure based in name/value pairs and lists. Illustrating this here is beyond the scope of this report, as it requires some technical knowledge. But JSON's simple structure means it may be used as a substitute for XML. XML is very verbose, and some files marked up in XML take a long time to process. The simplicity of JSON is designed to avoid that problem. Finally, as with XML, JSON has a schema language, which allows designers to construct reusable JSON elements and structures.

RDF/RDFS (Resource Description Framework)

The Resource Description Framework (RDF) was designed as a language for representing metadata about Web resources, or information about physical resources that can be identified on the Web. This framework is not meant to be particularly human readable, but rather is meant to work behind the scenes to facilitate search and retrieval (see Figure 5-2 for an example). Where XML is inherently document-like and hierarchical in its structure, RDF is a methodology for creating individual, simple, and interconnected statements about resources. Each individual statement may be represented as a graph—two nodes and an arc connecting them. So a set of many interconnected statements is representable as a complex graph. These statements are structured as subject-predicate-object triples (named for the trio of elements a triple contains). The subject is the entity being described; the predicate is the relationship between the subject and object; and the object is the value, entity, or other characteristic being ascribed to the subject. The object in one triple can be used as the subject of another triple, thus enabling a rich set of interconnected triples. In the graph representation of a single RDF triple, the nodes represent the subject and object, and the arc represents the predicate.

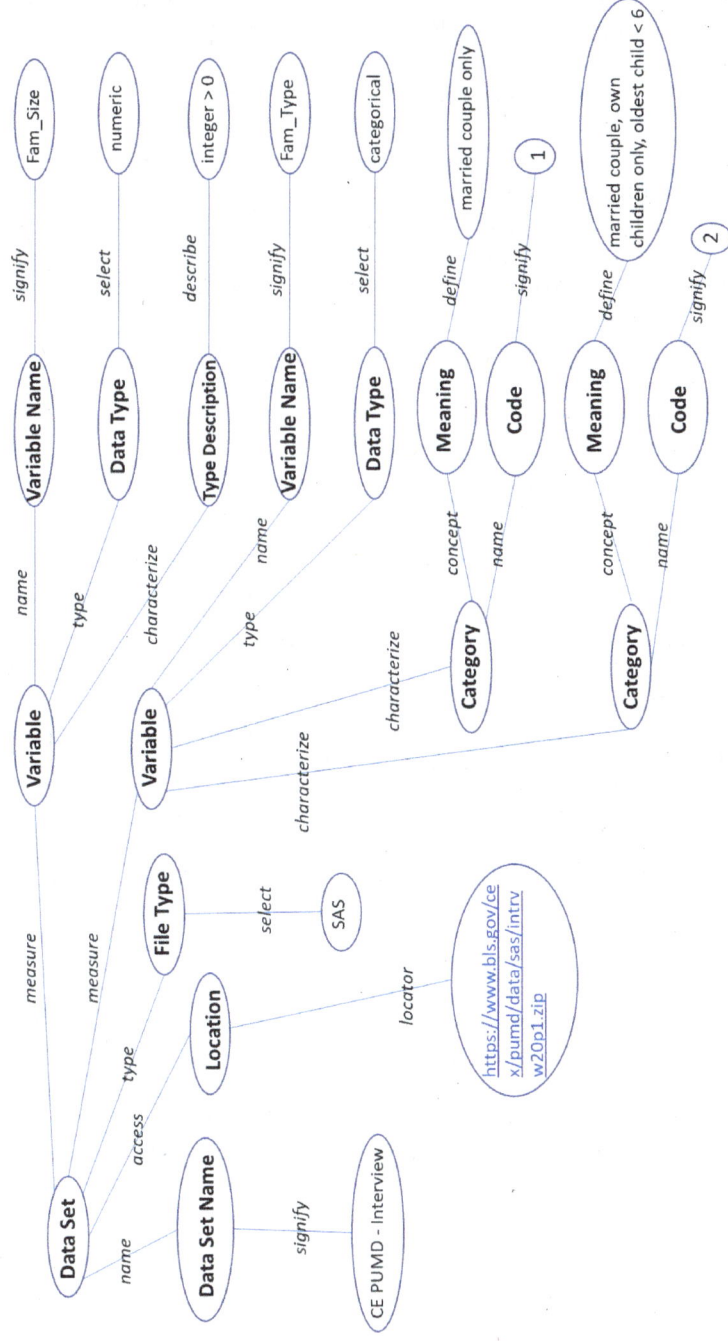

Figure 5-2 A simple dataset description in RDF.

The semantics (or meaning) of the subject, predicate, and object should all be designated by a Uniform Resource Identifier (URI), which is used to refer to a meaning or value. The ability to create an interconnected graph from individual statements depends on the consistent use and linking of URIs. Data formatted in RDF are often referred to as "Linked Data" because of this.

Just as with XML and JSON, there is a schema language for RDF, which is called RDFS (Resource Description Framework Schema). The schema allows designers to build an RDF structure that can be reused. This is important when using RDF to represent metadata describing similar resources.

RDF is increasingly used by institutions that preserve information (see, for example, the Library of Congress's Linked Data Service[6]). However, it is important to keep in mind that RDF is designed to be consumed by machines, not people. A graphic user interface is needed to resolve URIs into human-readable text about resources. Any adoption of RDF needs to be accompanied by suitable software for end users.

METADATA SYSTEMS[7]

Metadata systems are built to organize recorded metadata. They are designed to make use of metadata in some way, either to provide human-readable metadata for informational purposes or to provide machine-readable metadata for guiding further processing and possibly for humans to read. There are two main components of any metadata system:

- The *metadata repository* is a database of metadata, the basic storage component for all metadata systems. The organization of the repository is based on a schema (e.g., column headings, relationships, and data types) to help formalize and organize the attributes, or metadata elements, of the system. See Appendix A for more details.
- The *user interface* is the means, possibly in the form of software, for a user to interact with a metadata system. Users can be humans or other systems, and these correspond roughly to whether the metadata are human readable or machine readable, respectively. The interfaces will be built based on which kind of user they

---

[6]See https://id.loc.gov/ for details.

[7]Much of this section is taken from Scope Metadata Team, "Metadata Systems for the U.S. Statistical Agencies, in Plain Language," 10 July 2020. The SCOPE Metadata team is an informal, longstanding interagency group of U.S. federal statistical agency representatives. Authors here include Daniel Gillman (BLS), Kathryn McNamara (Census), Peter B. Meyer (BLS), Francisco Moris (NSF/NCSES), William Savino (Census), and Bruce Taylor (IES/NCES).

address. In each case, an API is the software designed to communicate with the repository, and the API gets its guidance, that is, what commands to send to the repository, from either a human or another system. The API sends a query desired by the user to the repository and returns metadata to the user.

The FAIR guidelines, mentioned earlier, are a generic set of principles for direction on how to make scientific data *Findable, Accessible, Interoperable*, and *Reusable*. These four goals are each subdivided into three or four principles (15 in total) that bear on data and metadata management and dissemination. Therefore, they need to be applied to the development of metadata systems, as these systems will be expected to support the FAIR guidelines in time. And, as we shall see, these principles support the efforts to make data and processes transparent and the results of these processes reproducible.

Metadata projects—that is, the effort to build metadata systems—are often framed by four considerations:

- The intended use of the system under development;
- The staff, contractors, capabilities, preferred methods, and software tools;
- The standards; and
- Other constraints.

As noted in the review of costs below (see Risks and Benefits), these projects involve staff from many areas—management, subject-matter experts, statisticians, information scientists, computer scientists, and IT specialists. These can be complex projects. Constraints include costs, time, functionality, scope, staff, resources, capabilities, and training. All are major factors in designing, building, and maintaining any system, and remain true for metadata systems. The novelty of a metadata system at a small agency, combined with uncertainty about its benefits, makes it difficult for the agency to be willing to use it.

Whether a metadata system is designed and built from scratch or the software for the system is borrowed or bought from another agency or a company, much development work is required. Collecting and storing metadata have additional requirements beyond just ingesting data into a database. Metadata are descriptive, so the need to make sure they describe the right objects, the descriptions are correct, and the descriptions are coherent with each other adds significant complexity. As with most system development and because of their novelty, metadata systems are best built iteratively, which includes the content in addition to the infrastructure (the software). Success is probable so long as the scope at each step is well defined and narrow; there is support from upper management; there is technical and subject-matter support; funding is adequate for the proposed goals at each stage; the user community is knowledgeable and supportive and can provide feedback; and there is significant planning. Nevertheless, the novelty of metadata systems presents a considerable risk, which is attenuated by the iterative approach.

What slows down deployment is collecting, organizing, linking, and checking the quality of the metadata in a dedicated system. Metadata management is most successful when metadata are collected as they are created. An example is the definition of some concept, say a new category in an existing classification. Where is that definition stored? Usually, the answer is that it is stored in some Word or Excel file, but that solution affords little in the way of good metadata management. If, on the other hand, the definition were added to some existing system that manages categories in statistical classifications, none of the work to collect the definition post hoc is needed. Building and integrating the right tools for interacting with systems is a fundamental piece of an effective system. And even though the number of vendors and open-source tools for the statistical metadata community is limited, good ones nevertheless exist and are designed to be integrated.

To get off the ground, the development of metadata systems requires some kind of business requirement to motivate the process, and this, in turn, provides the justification for stakeholders to commit time and resources. Most of the time, the collaborative process is slower than the management of organizational development schedules can accommodate. This is a risk that needs to be acknowledged up front. Losing stakeholders in the middle of a development often causes that work to get bogged down further. If a stakeholder insists on a time schedule, either everyone must agree that the end product will probably be less than originally planned, or a new development round must be agreed to. Given the twin goals of transparency and interoperability in and among the 13 principal U.S. federal statistical agencies, collaborative projects are very likely the right way to proceed.

In addition to all of the above, achieving success requires that management and technical staff be supportive. The support of top management, especially, is necessary for two simple reasons: (1) control of the budget, and (2) the ability to direct their staff to perform certain tasks.

All-encompassing metadata systems are sometimes called "cathedrals"; they usually fail because they are too complex, take too long to build, and are too ambitious. Committed supporters of metadata management often make this mistake. Despite the skeptical view of those who refer to this work as cathedral building, in fact the metaphor also applies to what is needed to make it succeed: an incremental approach that allows sufficient time. Expectations need to be reined in by making the scope manageable at each step. One cannot simply start building such a cathedral instantly; first, all the aspects of the standards and tools that will eventually need to be incorporated must be considered. New kinds of systems need to be built slowly, so while one cannot start assembling this cathedral initially or all at once, the broad vision needs to be kept in mind. The iterative approach means the ultimate design of the cathedral and what is achievable must be fluid. Yet all aspects of incorporating tools and conforming to standards need to be considered early on.

New kinds of systems, those that provide a new kind of functionality, need to be built slowly. This is especially true for metadata management, for all the reasons laid out above. Future iterations need to incorporate lessons learned from past steps. Few iterations will be mistake free, and often the best experience is one that involves some failure. The potential for metadata management is often not completely understood at first. This will most likely change as experience is gained through system development. The scope of a system may need modification as resources are re-examined midstream. Careful system development requires all these considerations, but they bear repeating as they are ignored easily. When new functionality is being considered, they are even more important.

RISKS AND BENEFITS

Given the costs and time that must be devoted to training and developing tools to adopt and make use of any of the six metadata standards described in detail later in this chapter, there is understandable hesitancy about building the capabilities of statistical metadata management. Some agencies view the making of informed use of metadata standards as "a bridge too far." However, the costs are not excessive, and the benefits will extend long into the future. Metadata standards are a key tool in minimizing burden and enabling transparency and future sharing of information and data products.

No system development will be approved by upper management unless the system is expected to provide a return on investment. Here we discuss some of the main areas where metadata systems may produce a return on investment, in which the risks are outweighed by the benefits.

Building and using metadata systems include the following transformative benefits:

- *Maintaining organizational knowledge.* An agency's knowledge is stored in documents and databases throughout the organization. These documents and data are metadata, because they describe something the agency planned, decided, built, used, or disseminated. Documents are usually written in prose and meant for humans to read. Metadata is a shorthand version of documents. By compressing the information contained in a document into attributes and their values, the metadata that result provide a shorthand for the prose. When stored in a repository, the metadata are made available for use by systems, human or otherwise. In this way, we see metadata systems supporting knowledge management and retention.

- *Improving user experiences.* A basic purpose of metadata systems is to help users find, understand, compare, and appropriately use data and other resources. Through the information contained in metadata, users are provided with a reference library that facilitates these activities. Metadata systems can also improve the public perception of an agency because they render data easier to use. User-provided metadata can also be added to systems, simplifying the experience of other users facing similar problems.
- *Improving producer experiences.* The production of statistics requires documentation for designs and processing steps. Increasingly, these are produced in a machine-readable form. In this form, metadata can be used as input parameters to processing systems, to trace or audit the development of output data, and to find ways to streamline and troubleshoot the production process.
- *Ability to be reused.* An advantage of machine-readable metadata is that they can be stored one time and reused. The Federal Data Strategy uses the phrase "Define once; use many times." Metadata are reused when they help describe many resources. For instance, a variable used in every dataset produced by a monthly survey only needs to be described one time. Similar variables used in several surveys share common attributes, and those attributes need to and can be reused. There are many more of these kinds of examples.
- *System development.* By using established standards, automatically interoperable datasets may be produced. The metadata systems provide the necessary metadata and structure on how the data are to be organized for users on release. Another application is the automatic production of a survey instrument from a structured set of questions, response choices, question flow, and interviewer interface, all of which are metadata from the instrument point of view.
- *Improvement to data governance.* Having a recognizable and maintainable set of metadata concepts that can be administered by accountable entities can be used to align different statistical projects and to provide a commonly agreed-upon conceptual framework that can improve transparency.

These benefits accrue both to the agency that adopts metadata systems and to the users of the data they produce. Further, they enable transparency in the agency's activities. Across agencies, metadata systems provide a common language for transparent communication.

Consider, now, the risks of metadata systems. Projects to create them involve staff from many areas—management, subject-matter experts, statisticians, information scientists, computer scientists, and IT specialists. Costs also involve staff capabilities and training. The novelty of such a system is one more thing that triggers resistance among staff to adopt it. Management and outside agency support could be important needed additions.

Next, planning for metadata management often necessitates a deep rethinking of how to organize information. Past inconsistencies, ambiguities, mistakes, and failures may be exposed, and these can be sensitive. Concerns also arise if too many people are found to be lacking in the skill or training needed for the tasks involved in creating a metadata system, which can seriously reduce management's willingness to move forward. Often goals are reined in, and projects may not always recover their original intent.

Metadata are best recorded when descriptions of resources are first devised (at "think time"). Often, however, metadata management is an afterthought in the planning and design stages of the statistical or survey life cycle, and this leads to the perception that collecting metadata *ex post facto* is expensive and time consuming. This problem repeats often, and metadata systems are built rarely. Metadata management functions are not in the original plans for many systems and statistical programs, so retrofitting metadata into a system built later does not happen easily. The problem may be addressed by always planning for metadata management activities in any system design. This is a key habit to form.

In short, the costs of adopting metadata standards and management procedures are varied and can be significant. They are lower, however, if adoption occurs prospectively, or if the work of retrofitting established systems is done iteratively. Most important, the benefits discussed next outweigh these costs, whenever and however they are incurred.

There is the reverse question as to whether there is a cost to *not* building metadata systems. This is rarely considered. The following is a list of advantages metadata management provides,[8] which easily translate into increased costs if these same advantages are not available:

1. *Better organization and search.* Metadata make it possible to find data, know where they come from, who owns them, what they mean, and how to use them responsibly.

---

[8] This list is taken from "The Value of Metadata," by Kurt Cagle, published on February 26, 2019, at https://www.forbes.com/sites/cognitiveworld/2019/the-value-of-metadata/?sh=2ba581e56d30.

2. *Reduced software costs.* Knowing what data mean and how they are structured makes it possible to write reusable, quality software more quickly and reliably.
3. *Richer data analytics.* Datasets often have missing or erroneous values, and these can automatically be identified by checking values against metadata.
4. *Easier governance.* Data governance focuses on the following correspondence to who, what, when, where, why, and how: authoritativeness (who is responsible for ensuring the integrity of that data), dominance (what constitutes a primary record), temporality (when were the data first known), provenance (where information comes from), purpose (why were the data captured in the first place), and definition (how exactly the information is defined, often including citational context).
5. *Understanding the external data environment.* Metadata are needed to integrate internally obtained data with those from outside sources.
6. *Making your organization machine readable.* When metadata are machine readable and actionable, this allows systems to speak a common language. In turn, this raises the value of the data the organization produces, because they can be integrated with other data more readily.

Each of these advantages becomes a cost when it is unattainable. Since each advantage depends on metadata being available, it is the existence of a metadata system that supports and affords the advantages. Not managing metadata is therefore a cost contributor.

For the National Center for Science and Engineering Statistics and U.S. statistical agencies to put into operation standards writ large, governance is required. The panel believes such a governance entity currently exists, namely the Interagency Council on Statistical Policy (ICSP).

Recommendation 5.1: The Interagency Council on Statistical Policy should develop and implement a multi-agency pilot project to explore and evaluate employing existing metadata standards and tools to accomplish data sharing, data access, and data reuse. The National Center for Science and Engineering Statistics should be an active agency participant in the project.

For this purpose, ICSP should form and monitor an interagency committee to design and implement a multi-agency project that benefits from the use of metadata standards and tools to increase transparency and to support interagency sharing of data and methods. The goal of the pilot project would be to explore the issues associated with improving the transparency and efficiency of data transfers across federal statistical agencies through the more widespread use of metadata standards and tools. The pilot would document these issues for dissemination to the ICSP agencies and obtain a broader understanding and potential of the use of metadata standards. Such a pilot project should serve as a foundation for additional follow-up projects. These projects should lead to statistical agencies developing policies on the use of metadata standards and tools in documenting methods and retaining for their input data and official estimates for future use.

USING EXISTING SYSTEMS

Instead of designing new systems from scratch, the use of metadata standards greatly increases the likelihood that existing systems can be shared, recycled, and repurposed. Indeed, the advantage of adapting existing systems is a primary reason to adopt standards, as this reduces the costs of a new development, reduces the implementation time, increases quality (the system software is already tested), and improves collaboration across the statistics community.

Using metadata standards for systems integration would allow one to move to a new paradigm, in which the following actions would be effective:

- Analyze existing metadata systems for integration rather than building new systems by default;
- Extend existing systems if they do not cover all of the required statistical life-cycle activities or functionality;
- Design new systems only when required, and design for potential reuse so that they may be adopted by others in the future, thereby creating a value-added feedback loop.

The principles in the Common Statistical Production Architecture[9] standard describe a framework for reusable systems architecture.

Here are some examples where metadata standards have enabled system sharing and reuse:

- The freely available SDMX tools,[10] developed and maintained by Eurostat to disseminate SDMX data, are implemented by many agencies worldwide, thus lowering the barriers of SDMX implementation within and across countries.

---

[9]https://statswiki.unece.org/pages/viewpage.action?pageId=112132835.
[10]https://ec.europa.eu/eurostat/web/sdmx-infospace/sdmx-it-tools.

- The International Household Survey Network (IHSN),[11] managed by the World Bank, is a program whereby guidance and tools are freely provided to developing countries using DDI-Codebook to help document their surveys and censuses.
- A statistical systems catalog[12] lists open-source systems mapped to the Generic Statistical Business Process Model (GSBPM), enabling an architect to find a system to integrate or extend for the required statistical activity and purpose.
- The Web page data.gov,[13] managed and hosted by the U.S. General Services Administration, provides tools and metadata schema for building and maintaining a dataset catalog, under the Resources tab on the project home page.
- The Web page schema.org[14] is run by a collaborative program whose mission is to create, maintain, and promote schemas for structuring and describing data, especially on the Web, but which can be applied widely. Any dataset whose metadata are in schema.org is identified as a dataset in Google data search.
- The open-source .Stat platform[15] is an SDMX-based and CSPA-inspired system for statistics data and metadata storage and dissemination, available from the Statistical Information System Collaboration Community (SIS-CC). It is used by several international agencies and organizations, enabling them to avoid having to develop such a platform themselves and instead benefitting from the collaborative developments from the SIS-CC community.
- Seasonal adjustment software JDemetra+[16] from the European commission is implemented by many EU and non-EU agencies, helping standardize seasonal adjustment processes and lower costs.

The following are four examples of how metadata standards have been used to benefit transparency in the provision of official statistics: at the Bureau of Labor Statistics (BLS), the World Bank, the U.S. Census Bureau, and the survey contractor Westat.

At the Bureau of Labor Statistics

The Consumer Expenditure Survey (CE) program at BLS is used to measure spending by households in the United States. The data are also

---

[11] https://ihsn.org/.
[12] https://github.com/SNStatComp/awesome-official-statistics-software.
[13] https://www.data.gov/about.
[14] https://schema.org/docs/documents.html.
[15] https://siscc.org/stat-suite/.
[16] https://ec.europa.eu/eurostat/cros/content/software-jdemetra_en.

used as input for calculating the Consumer Price Indexes. This short section is about the efforts at BLS to make CE more transparent.

CE is conducted through two ongoing surveys: the (weekly) Diary and (quarterly) Interview. Sampled households for the Diary survey are given a form to record small expenses for each of 2 consecutive weeks, and those in the Interview survey answer questions about large or recurring expenses for each of the previous 3 months for four consecutive quarters. The Interview survey is conducted every month on a rotating sample. The data are combined on a monthly basis to produce quarterly, semi-annual, and annual estimates and datasets. This includes the annual public-use microdata (PUMD) files. BLS posts these files on the CE Website along with a large (~120 pages) documentation file.

Achieving transparency for the PUMD requires more information than the annual documentation files can convey systematically. Over time, CE incorporates changes in the even years. The surveys collect some of the same data in different ways, so the provenance of the final data (the path from collection to dissemination) is complex. Nevertheless, all these considerations are necessary for understanding and using the PUMD, especially when using the data over time.

The CE office examined its transparency needs for the PUMD, for both the public and office analysts. A team was formed to recommend the development of systems built iteratively to demonstrate increased functionality toward the CE transparency goals. The team examined the DDI-Lifecycle standard (now version 3.3) from the DDI, and determined that it could satisfy their informational needs. Commercial software (Colectica) that would ensure conformance with the DDI-Lifecycle was purchased, and it was determined to be flexible enough to meet CE development needs. The current system is the fourth in the series.

Currently, this system is being moved to a production server to support internal CE analysts. The system contains mappings of all the variables to their respective questions and mappings of questions to the variables, satisfying the basic need to describe the data over time and across surveys. Another system is planned to make substantially the same information available to the public. Further development is planned as well, especially to describe all the processing steps (edits, coding, allocation, and imputation) used to transform the data after collection. Here is where the main differences between the internal and public systems lie: a public system must hide processing details, because some algorithms could be reversed to create disclosures.

Future planned work includes expanding the description of processing, especially for internal use, and improving the user interface to facilitate use among the public.

The idea for this project emerged as analysts' needs were assessed and a middle manager saw an opportunity. Since each step of development has been relatively small, risks have been minimized. Upper management has been willing to fund the iterative approach, one step at a time. While this slows progress, because there is then the need to create new statements of work, it provides time for upper management to demonstrate progress and new functionality. Such an iterative approach is often conducive to ensuring progress, because the goals are small each step of the way. An added advantage is the ability to change any long-range vision as lessons are learned along the way. Each phase of development was carefully constrained so development stayed within the stated goals. The current development is in the fourth phase, and the success at each phase has increased management support for the project.

At the World Bank

The World Bank maintains a Microdata Library, a catalog of more than 10,000 survey datasets originating from member countries and from regional and international organizations. To maximize the discoverability and usability of the data, the World Bank documents these datasets using the DDI Codebook and the Dublin Core metadata standards. A specialized software application for metadata editing facilitates this process, allowing fast documentation of the surveys, their data files, the variables they contain, and all related resources (questionnaires, manuals, reports, and others). An open-source cataloging application is used to catalog and disseminate the data and metadata. The DDI Codebook and Dublin Core metadata are stored as XML and JSON files. These formats ensure machine readability and offer all the flexibility needed to feed meta-databases and to generate human-readable output, including PDF documents, HTML Web pages, and others. The DDI Codebook and Dublin Core metadata provide rich, structured metadata that cover all dimensions of a data collection operation: objectives, concepts, methods, scope and coverage, universe, sampling methods, processing methods (editing and imputation methods), quality and relationships to other datasets, and a detailed data dictionary.

The result is transparency concerning *what, how, why, when,* and *who*. Metadata produced by data curators can be "augmented" using various machine learning techniques such as topic models, word embeddings, and image labeling. This information allows users to assess multiple dimensions of data quality for their specific purpose, including relevance, reliability, usability, and comparability. For example, the cataloging of variable-level information allows users of the World Bank Microdata Library to locate variables of interest in multiple datasets, then assess their consistency and comparability across sources and/or over time.

The World Bank is now expanding the scope of its cataloging system to other types of data, for which other metadata standards are used (such as the ISO19139 standard for geospatial data, and a custom-designed schema to document reproducible analytical scripts). The combination of rich and structured metadata with advanced indexing and machine learning tools is expected to result soon in significant progress in data discoverability and transparency, including semantic searchability and recommender systems. And the metadata schemas are expected to evolve as new data collection modes and new data sources become more prevalent.

At the U.S. Census Bureau

A group of analysts at the Census Bureau are using metadata standards for data discovery. This is because their group makes use of administrative data that are often acquired and then rarely used because their existence is difficult to know about. Having a searchable metadata catalog was the solution, but their existing data management system did not have much in the way of file or variable metadata, so they needed to harvest such metadata directly from the data files that had been discovered (half a million data files from hundreds of data families). To collect these metadata, they created a data discovery interface to identify useful data files, whose metadata were then harvested. This enabled others to find and use these data files for their programs.

The Census Bureau believes it should make greater use of DDI Lifecycle and that the agencies within the U.S. federal statistical system should have greater participation in the development of DDI standards, other standards, and metadata tools than they currently do. This is because of the concepts DDI Lifecycle uses to represent variables, the file/table linkage concepts that support record linkage, and the above-noted use for multiyear studies. Also, use of DDI facilitates the transfer of metadata outside of an agency.

At Westat

Westat has had a contract for several data collection cycles with the Agency for Healthcare Research and Quality (AHRQ) to collect data for the Medical Expenditure Panel Survey (MEPS). Westat provides about 4,000 variables (many derived) to AHRQ, with about 200 deliverables annually stemming from the data collected. A major technical upgrade to the household component of MEPS (there are also provider and insurer components) was viewed as necessary due to the lack of flexibility of the data collection system, which had been in place for more than a decade. This lack of flexibility constrained access and the sharing of data and metadata. Part of the solution was a custom build-out of a new metadata system, using commercial off-the-shelf software. After reviewing several alternatives, Westat decided to use DDI for this purpose. Its advantages were threefold: (1) it was already familiar to a large number of users from multiple disciplines; (2) multiple data products already used this metadata standard; and (3) it was based on a common programming language, XML.

Algenta Technologies, a research and development firm specializing in data management, assisted Westat in carrying out this implementation, using its Colectica software. Among the reasons for deciding to use Colectica were that it used an SQL[17] database to store metadata in XML-based DDI 3.2, there was a high level of support from the company for this custom work, and it can be integrated with Blaise.[18]

At the same time, MEPS designed and implemented portions of a complex repository for the data and the metadata. The repository supports tracking questions and variables over time. Access is available either through a Web portal or through MEPS project-specific repository tools. As a result of these changes, MEPS has improved in the following ways: it is now DDI compliant, which makes it easier to access by users and makes its contents easier to understand; it supports a graphic user interface; it supports Blaise integration; it supports custom outputs; and it is designed to maintain data lineage.[19]

STANDARDS AND INTEROPERABILITY

This section discusses standards and whether they are based on reusable schemas. Standards are effective when they define schemas that are reusable across a range of applications. The range of applications, which is known as the scope of a standard, determines whether the underlying schema in a standard might apply.

Adopting standards is a means by which the separate U.S. federal statistical agencies can achieve some uniformity and interoperability among their data and metadata systems in terms of both data stores and services. When left to their own devices, the agencies build systems in their own ways. An example of the effect of this is the variation among the APIs that many of the agencies have built and promote. These APIs are designed for people wanting to access data programmatically from individual agencies, but their variation exacerbates differences—such as the differences in how the agencies format dates—and the APIs provide little metadata. This leads to incompatibilities that, given their multiplicity, are costly to overcome, since every pair of APIs has to be harmonized (13 principal statistical agencies

---

[17]SQL is Structured Query Language, described at https://www.infoworld.com/article/3219795/what-is-sql-the-first-language-of-data-analysis.html.

[18]https://blaise.com/blaise/about-blaise.

[19]For those interested in a non-U.S. example, please see the following on the Aristotle Metadata Registry: https://www.acspri.org.au/iassist2019/workshopAristotle.

leads to 78 pair-wise translations). This also leads to the inertia the U.S. federal statistical system finds itself fighting in the effort to overcome the differences the individual approach has fostered.

Participation in Standards Development

Standards are specifications designed to solve the problems described above and, in general, to satisfy the business requirements of stakeholders (those organizations with a material interest in the outcome of the standards development process) built through a *consensus*[20] process where

- Each stakeholder is afforded the same rights as any of the others—the process is *fair*;
- Any interested party may examine the results of the development process at any time—the process is *observable*;
- All stakeholders are eligible to join in the development of the standard—the process is *open*; and
- The stakeholders represent the broader community of potential users of the standard—the process is *balanced*.

In this report, we will refer to a standards development process that satisfies the fair, observable, open, and balanced criteria and the standards that result from those as *equitable*. The openness criterion in an equitable standards development process means that any stakeholder can join the development effort at any time. Since standards are not static, they undergo periodic review to check for continued relevance. They can be updated to fix problems, expanded to address new requirements, held static, reduced because some features are no longer relevant, or dropped altogether. Therefore, stakeholders are encouraged to join groups developing and maintaining standards at any point in the development life cycle. Joining a development effort does not require participation from the beginning.

Usually, a standard is better when it meets more stakeholders' needs, and this happens if more participate in its development. More participation means the chances all requirements are heard go up and are put on the table for consideration. This is what makes a standard useful for stakeholders. And the more useful a standard is, the better the chance it will be adopted.

On the other hand, each stakeholder has requirements that are unique, even though they might share most of their other requirements with the others. It is these unique requirements that sometimes lead to a decision not

---

[20]Consensus, as used here, means general agreement with no sustained dissent. Consensus is possible even when some stakeholders just agree to go along, rather than raise dissent. This is the difference that distinguishes consensus from unanimity.

to adopt a standard if the standard does not meet the needs. Those unique requirements, moreover, can only be added if the stakeholder joins in the development process. If some important requirement is not included, the stakeholder will not adopt the standard, and this could lead to a business disadvantage when other similar organizations do adopt it.

The U.S. federal statistical agencies have many requirements in common, and these requirements are mostly shared with other statistical offices around the world as well. Therefore, the incentive for any agency to join a standards development effort might be lowered if that agency is aware that another agency is already involved. Failing to get involved because of this would not be the best practice, since the unique requirements are much less likely to be included, reducing the chances that the standard will evolve to meet their needs, as shown in the paragraph above, but in the current environment with tight budgets, it is still a strong possibility.

The U.S. federal statistical agencies are not business competitors, so for agencies with smaller staffs, it might be possible to let other, larger agencies develop the standards, which the smaller agencies could then adopt. In that case, it would be necessary to work with staff from a larger agency who are involved. In this way, any special requirements could still be heard about among the standards developers, though the arrangement is not optimal.

Employing Equitable Standards

For the U.S. federal statistical community, developing or using equitable standards should be the goal. Equitable standards are more likely to be adopted by agencies that were not directly involved in their development, because equitable standards stand a better chance of meeting the needs of those agencies. To identify sources of equitable standards, one must first know which standards development organization adhere to the principles for equitable standards given above.

Adopting standards across the agencies would make translating among the data and metadata from each agency simpler. It would mean the agency systems are already "speaking" the same language. For the user of U.S. federal statistical data, finding, understanding, and using (in particular, integrating) data from two or more agencies would be simplified. For example, if agencies report dates in the same format, those data would not have to be examined and translated. A computer could compare the dates with no extra work. If agencies all used the term *frequency*, for example, to name the idea of the interval between the times when an ongoing survey is conducted or an indicator is published, then a computer could be programmed to look for that attribute when it was set up to do comparisons.

Interoperability is the condition under which data and metadata are translatable across agencies. In general, interoperability means that a receiver (user) of data (or metadata) can use those data without needing human intervention with the sender. There are several ways to subdivide the idea of interoperability, but for simplicity, we can describe the world as divided into syntactic and semantic interoperability. We illustrate this simply by using the examples above. If each agency uses the same format to transfer and publish dates, that is an example of syntactic interoperability. If each agency uses the same name for "frequency" (as described above), that is semantic interoperability. There are deeper considerations around this idea as well.

The adoption of standards is a means to achieve interoperability. However, what does it mean to adopt a standard? Often, in trying to be more precise, people refer to compliance as a way to express this. However, there is a precisely defined term of art, *conformance*, used by the ISO community. We will use this in what follows (and in Appendix B as well).

Standards are built using precise language called *provisions*. The term *provision* is defined in ISO/IEC Guide 2 and is illustrated in Appendix B, but we condense those discussions here for brevity. *Provision* and closely related terms can be described as follows:

- *Provision*: statement, instruction, recommendation, or requirement;
- *Statement*: expression that conveys information;
- *Instruction*: expression that conveys an action to be performed;
- *Recommendation*: expression that conveys advice or guidance;
- *Requirement*: expression that conveys criteria to be fulfilled.

Provisions are distinguished by the form of wording used to express them; instructions are expressed in the imperative mood, recommendations by the use of the auxiliary "should," and requirements by the use of the auxiliary "shall." Definitions, for instance, are statements, and they are often not written in sentences, so the form of statements varies. Conformance can thus be defined as satisfaction of all requirements. Satisfying a requirement might include needing to address some recommendations, statements, and instructions, too. In any case, this means that if a system satisfies all the requirements in a standard, it conforms to that standard. Again, the interested reader can consult Appendix B for more details and a worked example.

Our interest is interoperability, most of all, and this is why conformance to standards is useful. Standards can be developed for all kinds of purposes. Consider the colors used in traffic lights: red means *stop*, green means *go*, and amber means *the light is about to turn red*. These standard colors are used throughout the world in the same way. This is semantic interoperability. Revisiting the examples used earlier concerning date formats and the names used for frequency attributes, conformance to a standard helps establish interoperability. So, one could establish a standard for how dates are formatted.

In fact, date format standards already exist: ISO 8601 (Data elements and interchange formats: Information interchange) and ANSI/INCITS.30 (Representation of Calendar Date and Ordinal Date for Information Interchange). The ISO standard is less useful since it specifies a number of formats. The ANSI standard provides for a single format, one of the many in ISO 8601. Concerning frequency attributes, statistical metadata standards, such as the United Nations Economic Commission for Europe (UNECE) Generic Statistical Information Model and the DDI-Lifecycle standard, do specify the attributes necessary to record metadata about frequency.

When a system has statements claiming conformance to some standards around its use, the receiver of data from that system already knows what to expect. The receiving agency can build its own system and know in advance it will be able to automatically read and understand the sender's data. That is interoperability based on conformance.

Conformance to standards simplifies the translation problem for exchanging and understanding data and metadata between agencies. As discussed briefly above, there are 13 principal U.S. federal statistical agencies. If each agency does its own development, ignoring any agreements or standards the agencies could share, a translation between every pair of agencies must be built to make data and metadata interchangeable. There are "13 choose 2" or 78 such pairings (or combinations), so producing standards this way would be prohibitively costly and time consuming. Worse, if one agency upgrades its systems, under that approach the translations would all have to be redone. In fact, agencies do upgrade systems fairly often.

See Figure 5-3 for an illustration of the differences in the number of translations needed for the status quo with pair-wise agency agreements in place versus the standards-based approach.

In our setting, building metadata systems through the adoption of consensus metadata standards has several advantages:

- Developers do not have to spend time modeling the necessary metadata, as this thought process will have already been undertaken.
- Conformance allows for interoperability, as the above makes clear, and this eliminates the need to translate between pairs of systems.
- Conformance occurs at the systems interface (either an API or a human user interface), so each organization can optimize its systems for local needs.
- Transparency of data and systems and reproducibility of results depend on metadata being available. The metadata needed are based on some schema defining the necessary attributes, and this

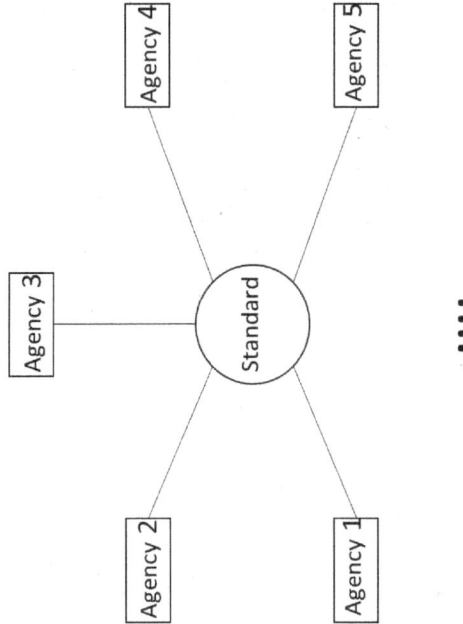
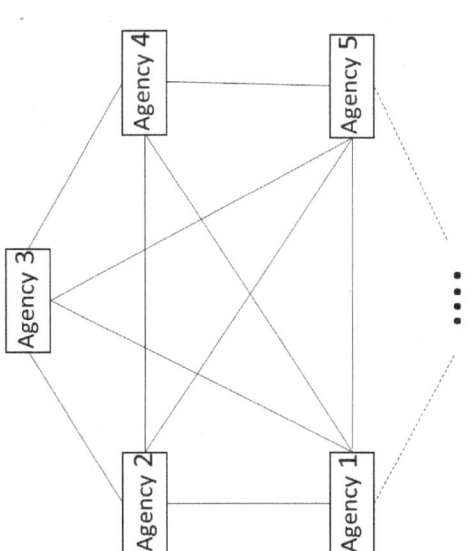

Figure 5-3 Conforming to standards—efficiencies gained.

can be achieved by adopting standards that specify those schemas and attributes.

EXAMPLES OF STATISTICAL METADATA STANDARDS

In this section, a brief description is given of six well-known metadata standards for statistics, including the main benefits each one has to offer. These six are the Generic Statistical Business Process Model. Generic Statistical Information Model. Common Statistical Production Architecture. Common Statistical Data Architecture. Data Documentation Initiative. and Statistical Data and Metadata eXchange.

Generic Statistical Business Process Model

The GSBPM is a matrix containing names and definitions for the business activities a statistical agency might employ to produce its data. It was produced under the auspices of the statistical program of UNECE. Even though the title of this standard includes the words "process model," the standard does not describe processes. Instead, it names the activities that processes are built to realize.

This standard is intended to be interpreted and used flexibly. It specifies the possible steps in a statistical business life cycle. The framework is presented in a particular order, but this is not rigid, and the items may be followed in any order. In fact, some steps may even be repeated in iterative applications, such as with some machine learning applications or imputation methods. Viewed as a checklist, GSBPM may be used to ensure that all necessary steps have been considered for conducting some statistical activity.

GSBPM is very flexible and can be used in a myriad of ways. The most direct kind of usage is to take the set of phases and activities literally, and use the standard as a guide on how to conduct surveys and other statistical studies. This might not be useful, as the language used in GSBPM may not correspond to the terms used in each agency. While the Census Bureau and the BLS have both adopted GSBPM, neither is using the standard as is. Instead, each agency has adapted it to the particulars of the work of the agencies and the terms in use there.

GSBPM can also be used for classifying systems, software purchases, and development projects, which all represent important use cases. Every statistical agency develops software and builds systems for processing the data produced through surveys and other statistical programs. The systems and the projects to build them can be classified according to the phases and sub-activities in GSBPM. In fact, this is the main application of GSBPM for the BLS and Census profiles (adaptations).

Generic Statistical Information Model

The GSIM is an internationally endorsed reference framework for information describing statistical programs, processing, and data. It was developed under the auspices of the UNECE as a companion standard to GSBPM. This generic conceptual framework is designed to support modernizing, streamlining, and aligning the work of statistical offices, such as the principal U.S. federal statistical agencies, and is one of the building blocks for modernizing official statistics.

GSIM is a conceptual model of the kinds of objects—surveys, questionnaires, variables, datasets—needed to describe statistical programs, processing, and data. The model contains classes (kinds of objects) and relationships that identify, describe, and relate these objects. Each object corresponding to a class in GSIM is informational, that is, each one describes things statistical agencies need to manage. The data associated with these objects are metadata, and therefore GSIM is a statistical metadata standard.

There is a deep relationship between GSIM and GSBPM. The inputs and outputs of the processes implementing the activities of GSBPM are the classes in GSIM. This *duality* between GSIM and GSBPM is characteristic of the relationship between process and information models. In practical terms, it means that GSIM and GSBPM are designed to work together, and agencies that adopt both gain more than through the adoption of GSBPM or GSIM on its own.

Another effect of implementing GSIM in a statistical office is the possibility of discarding subject-matter silos, due to the organizationwide view that GSIM promotes. Similarities among objects from several statistical programs, rather than differences, may be highlighted through the use of GSIM. This is an advantage of corporate (agencywide) metadata management, which differs from metadata management implemented for each statistical program independently.

Common Statistical Production Architecture

The Common Statistical Production Architecture (CSPA) developed under UNECE is a reference architecture for the processing required to produce statistical data. Whereas GSBPM lays out the activities a statistical office might implement, CSPA calls for the processes implementing those activities to be built as services that any agency can use. This goal is very ambitious, but CSPA can be implemented within a statistical agency by building common services across business silos, such as statistical programs and agency divisions.

The main vision of CSPA is to enable the development of "plug and play" statistical services, which will increase the reuse of systems, thereby reducing costs and increasing the speed of implementation. This is in contrast to the traditional development of stovepipe systems, each designed for a particular statistical program or within an agency silo, with the same kinds of services developed many times throughout a statistical agency (and the community at large) to solve the same or very similar problems. A simple example might be the need to build separate top-coding edit services for each statistical program that needs one. From a conceptual point of view, top-coding is a similar operation wherever it is used. Therefore, an important goal of CSPA is that the resulting services are interoperable with other CSPA services, and this means generically built services can work with others. The inputs and outputs of these services will be describable in GSIM, and it is possible for some of the outputs of one service to become the inputs of another.

CSPA consists of application and technology architectures and principles for the delivery of statistical services. One way to think of CSPA is as a service-oriented architecture that has been specially adapted for statistical activities and processes. It is not prescriptive of specific technologies or platforms.

CSPA can be combined with the GSBPM and GSIM frameworks. For example, if an agency has adopted GSBPM, and then has determined that a service is required for a GSBPM activity, then either

- the agency may develop a CSPA service, described in terms of the GSBPM activities it is used for, and other agencies may reuse it for those same activities; or
- the agency may identify and reuse a CSPA service already developed by another agency that matches the same GSBPM activity (a CSPA Global Artefact Catalogue exists for such services).

However, an agency may still benefit from the CSPA principles without fully implementing GSBPM or GSIM, especially to move toward a service-oriented architecture for statistical systems.

Common Statistical Data Architecture

The Common Statistical Data Architecture (CSDA), developed under the UNECE, is a reference architecture and set of guidelines for managing statistics data and metadata throughout the statistical life cycle. The latest version was published in 2018. As part of the family of UNECE standards, it has the same maintenance and governance administration as GSBPM and GSIM.

The purpose and use of CSDA is to act as a template for statistical agencies in the development of their own enterprise data architectures. In turn, this will guide solution architects and builders in the development of systems that support the production of statistical products. Because data need metadata to be useful, CSDA covers both data and metadata, and it refers to them, taken together, as "information." CSDA considers that the physical location of data does not matter (e.g., consider cloud storage), and in any case, the same principles apply. CSDA stresses that statistical information should be treated as an asset.

The benefits of using CSDA are a greater independence from technology, by being less dependent on proprietary software; greater sustainability, through use of a shared vocabulary; greater maintainability, through the shared architecture; and a cost-saving for global optimization strategies/solutions, thanks to sharing resources and technologies.

The CSDA standard consists of a set of key principles. These principles, their rationale, and their implications are described in Appendix A.

Data Documentation Initiative

The DDI, is a family of statistical metadata standards and other work products developed, maintained, and supported under a consortium called the DDI Alliance. This is managed through a secretariat at the Inter-university Consortium for Political and Social Research (ICPSR) at the University of Michigan. Below is a short description of the major standards, either published or in substantial draft form.

DDI2: Codebook

DDI2: Codebook, currently version 2.5, is used to describe a social, behavioral, or economic (SBE) research study, a one-time survey or experiment, or the data each might produce. There are many applications of Codebook in use. The data archive managed by ICPSR and the International Household Survey Network managed by the World Bank both use Codebook as their underlying metadata model. However, Codebook does not support reuse of metadata, so any interconnections have to be maintained outside Codebook itself. This feature makes it easy to implement but hard to discover inter-relationships. Codebook is sometimes used as the initial step in developing metadata management capabilities in the statistical domain. Codebook is managed in a directly implementable form in XML.[21]

---

[21]XML is eXtensible Markup Language, described at https://www.w3.org/XML/. See https://ddialliance.org/Specification/DDI-Codebook/2.5/ for more details.

DDI3: Lifecycle

DDI3: Lifecycle, currently version 3.3, is used to describe the activities in the statistical life cycle for statistical programs, such as censuses and surveys. The statistical life cycle used in DDI3 is consistent with the phases of the GSBPM, and this is consistent with the work in U.S. federal statistical agencies. Reuse is a necessary part of the design of Lifecycle, because most high-profile surveys conducted by U.S. federal statistical agencies are ongoing. More statistical offices in the United States and around the world are adopting DDI3 for their metadata needs, including the Bureau of Labor Statistics, Statistics Canada, and the Australian Bureau of Statistics. Lifecycle is also managed in XML and is directly implementable.[22] Because DDI3 supports GSBPM and is represented in XML, it is both consistent with GSIM and may be used as a physical (or implementable) manifestation of it. There is no need to build a bespoke implementation model for GSIM.

DDI4: Cross-Domain Integration

DDI4: Cross-Domain Integration (DDI-CDI), is the newest entry in the collection of DDI standards. A draft was released for public review and comments in April 2020. The final release of DDI-CDI is planned for the latter half of 2021. A new feature in DDI-CDI is the ability to describe data in four logical data formats. In addition to traditional survey data (microdata and multidimensional data), sources include administrative, remote sensor, Web scraping, and Internet streaming. Another feature is a generalized process model, which is used to describe how data are processed or produced, which includes the provenance of data. An innovation is the datum-centered approach. This allows any datum to be tracked through changes in representation, transformations during processing, and movement from one data structure to another. Metadata reuse is supported, and an infrastructure for integrating data from multiple sources is provided. The specification is built as a UML (Unified Modeling Language) model, and this facilitates easy maintenance. Multiple language representations can be generated, such as XML (which already exists), RDF (in production), and SQL (planned[23]). DDI-CDI has functionality that DDI3 lacks, and the two may be used together, but DDI-CDI was developed for independent purposes.

---

[22] See http://www.ddialliance.org/Specification/DDI-Lifecycle/3.3/.
[23] https://ddi-alliance.atlassian.net/wiki/spaces/DDI4/pages/491703/Moving+Forward+Project+DDI4.

Statistical Data and Metadata eXchange

The Statistical Data and Metadata eXchange (SDMX) was developed by a group of international statistical offices and banks. Their need was to produce a standard exchange system for reporting data and metadata to them from data-producing organizations. This group comprises the Bank of International Settlements, the European Central Bank, Eurostat (the statistical office of the European Union), the International Monetary Fund (IMF), the Organisation for Economic Co-operation and Development, the United Nations Statistics Division, and the World Bank. Since the publication of the initial version in 2004, the number of implementations continues to grow, covering many domains in statistics. These include education, labor, national accounts, and sustainable development goals. The current revision work being done on the standard will improve its capability for the exchange of statistical information.

SDMX consists of three main elements:

- Technical standards (including the Information Model),
- Statistical guidelines (sometimes called Content Oriented Guidelines), and
- IT architecture and tools.

These three elements can be implemented in a stepwise approach.

SDMX can support various use cases and implementation scenarios. Whatever the scenario envisaged, the essence of SDMX is to provide a set of common tools, processes, terminologies, and methodologies facilitating the exchange of multidimensional statistical data and metadata between producers and consumers.

The typical use case for SDMX is reporting data and metadata from a data producer to an international organization under some mandate or agreement. For example, all countries report national indicators to the IMF via SDMX to the Dissemination Standards Bulletin Board at https://dsbb.imf.org/. The infrastructure defined and used by SDMX to achieve this reporting objective contains a network of registries that make specialized SDMX structures, called Data Structure Definitions (DSDs), publicly and centrally available. A DSD contains a description of the dimensions and measures used in multidimensional data. Additionally, there is a set of reporting formats as well as various IT tools to assist implementers. Full implementation of such infrastructure will support direct machine-to-machine interchange.

Another area SDMX supports is data discovery and visualization. In this case the objective pursued by a statistical organization is to make statistical data and metadata findable, understandable, and accessible by external users. Data and metadata managed under SDMX are machine actionable, which means that data can be formally queried, not just indexed by keywords. The result is the ability to use many visualization tools with SDMX data. Another advantage to machine readability of SDMX data is for ETL (extract, transfer, and load) operations. These operations can be automated, providing many advantages.

Metadata and the U.S. Federal Statistical System

The decentralization of the U.S. federal statistical system increases the need for metadata, generally, and standards in particular. Metadata constitute the information needed to understand statistical data, designs, and processing. Since the U.S. federal statistical system comprises many agencies, the interoperability of those metadata is a necessary precursor to the goal of systemwide transparency and reproducibility. Data users should not have to translate metadata from one agency to the next, as this is an impediment to transparency. Finally, substantial changes in the kinds of work that statistical agencies perform may flow from the adoption of metadata standards and management, as staffing based on current work processes may be adjusted and streamlined.

There is now the opportunity to avoid this problem. The possibility of a common set of metadata standards that all or most statistical agencies share is discussed in *Metadata Systems for the U.S. Statistical Agencies, in Plain Language*.[24]

Threats impelling the adoption of our metadata standards and management recommendations are the flip side of that opportunity. Government-wide initiatives on transparency and reproducibility, including the Evidence-Based Policy Act, may be at risk if agencies do not coalesce on common frameworks for documenting and sharing data. The constraints of current data production and documentation and sharing procedures threaten to make it impossible for agencies to take on important, forward-looking work that is currently out of its reach.

CONCLUSION

Use of a standard framework or process model such as GSBPM allows users of official statistics to match data products with processes in a way that would allow them to better understand exactly how those products came to be. Use of information and data models such as GSIM and CSDA further enhance this link between data and business processes by

---

[24]See https://nces.edu.gov/fcsm/pdf/Metadata.projects.plain.US.federal.statistics.pdf.

establishing standardized concepts and information objects that map between conceptual framework and implementation.

Adoption of standards such as DDI or SDMX is worthwhile, not only because transparency and reproducibility are inherently desirable as qualities that enhance the trustworthiness and reliability of statistical products and processes, but also because these standards were developed with the intention of helping national statistical agencies do more with less in a way that is interoperable and mutually intelligible. While it is difficult to accurately forecast issues that will arise in the long term, it is safe to say that the budgets and resources available to statistical organizations are not going to scale as quickly as the amount of data being produced will. Nor will they scale as quickly as the demand for quantitative evidence to inform policy addressing the myriad issues arising in our increasingly complex world. The case for implementing shared systems, models, and standards is further strengthened when considering that novel problems requiring statistical production may well be transnational or global in scope, and time spent attempting to translate between idiosyncratic terminology, processes, and practices of different nations' statistical institutions will amount to wasted time, effort, and resources.

Recommendation 5.2: The Interagency Council on Statistical Policy should (1) prioritize and emphasize the importance and benefits of federal statistical agency staff engaging in international metadata standards and tool development, and (2) organize a discussion among statistical agencies that leads to an effective, coordinated, and accountable approach for staff in agencies that produce federal statistics to contribute to international metadata standards and tool development.

A number of federal government mandates have been issued during the past few years, including the Federal Data Strategy, U.S. Office of Management and Budget circulars, the Evidence Act, and others. One theme common to all of them is the need for transparency in statistical operation, methods, and analysis. It is not obvious that agency staff are aware of the content and requirements found in these directives.

Recommendation 5.3: The Interagency Council on Statistical Policy and the Chief Statistician of the United States should develop a continuing education program for agency staff on the nature and purpose of federal government requirements for statistical activities (such as the Foundations for Evidence-Based Policymaking Act and the Federal Data Strategy) that have been issued, the expected positive return to the agency for their implementation, and the need for transparency in agencies that produce federal statistics.

Individual agency staff should attend and participate in such continuing education programs with the goals of gaining professional familiarity with metadata standards and tools and improving the transparency of statistical methods, operations, and analysis in their data programs.

6

Making the Practices of the National Center for Science and Engineering Statistics More Transparent

DESCRIPTION OF NCSES PROGRAMS

The National Center for Science and Engineering Statistics (NCSES), one of the 13 principal federal statistical agencies, is located within the Social, Behavioral, and Economic Sciences Directorate within the National Science Foundation. NCSES's goals are the collection, interpretation, analysis, and dissemination of objective data on the science and engineering enterprise. The center is responsible for collecting data and publishing official statistics on (1) research and development (R&D); (2) the science and engineering workforce; (3) U.S. competitiveness in science, engineering, technology, and research and development; and (4) the condition and progress of science, technology, engineering, and mathematics education in the United States. NCSES fulfills its mandate primarily through the collection, analysis, and dissemination of statistical data in support of research on these topics.

NCSES's survey portfolio includes 15 ongoing surveys and one upcoming survey (see Table 6-1). Of these ongoing surveys, 10 are annual, 4 are biennial, and the periodicity for one is to be determined. Ten of the ongoing surveys are performed under competitive commercial contract, and the remaining five are performed under negotiated interagency agreements with the Census Bureau.

In addition to data tables and other data products (see below), the primary survey outputs include several types of analytical reports. These include

TABLE 6-1 NCSES' Survey Portfolio

Name (Acronym)	Who Collects the Data / Type	Where Documentation is Available
Annual Business Survey (ABS)	Census Bureau/ sample	https://www.nsf.gov/statistics/srvyabs/ https://www.census.gov/programs-surveys/abs/about.html; https://www.census.gov/data/developers/data-sets/abs.html; https://www.census.gov/programs-surveys/abs/technical-documentation.html;
Business Enterprise Research and Development Survey (BERD)	Census Bureau/ sample	https://www.nsf.gov/statistics/srvyberd/; https://www.nsf.gov/statistics/srvyberd/#rSR&rOTH&tabs-1&sd https://www.census.gov/programs-surveys/brds/about.html;
Early Career Doctorates Survey (ECDS)	RTI International/ sample	https://www.nsf.gov/statistics/srvyecd/; https://www.nsf.gov/statistics/srvyecd/#sd; https://edsurveys.rti.org/ecds_ft/
FRDC Research and Development Survey (FFRDC R&D)	ICF/ census	https://www.nsf.gov/statistics/srvyffrdc/; https://www.nsf.gov/statistics/srvyffrdc/#sd;
Higher Education Research and Development Survey (HERD)	ICF/ census	https://www.nsf.gov/statistics/srvyherd/; https://www.nsf.gov/statistics/srvyherd/#sd
National Survey of College Graduates (NSCG)	Census Bureau/ sample	https://www.nsf.gov/statistics/srvygrads/; https://www.nsf.gov/statistics/srvygrads/#sd; https://www.census.gov/programs-surveys/nscg.html
Nonprofit Research Activities Survey (becoming a module of the ABS)	Census Bureau/ sample	https://www.nsf.gov/statistics/srvynpra/; https://www.nsf.gov/statistics/srvynpra/#sd;

TABLE 6-1 Continued

Name (Acronym)	Who Collects the Data / Type	Where Documentation is Available
Survey of Doctorate Recipients (SDR)	Westat/ sample	https://www.nsf.gov/statistics/srvydoctoratework/; https://www.nsf.gov/statistics/srvydoctoratework/#sd;
Survey of Earned Doctorates (SED)	RTI International/ census	https://www.nsf.gov/statistics/srvydoctorates/; https://www.nsf.gov/statistics/srvydoctorates/#sd;
Survey of Federal Funds for Research and Development (Fed Funds)	Synectics/ census	https://www.nsf.gov/statistics/srvyfedfunds/; https://www.nsf.gov/statistics/srvyfedfunds/#sd;
Survey of Federal Science and Engineering Support to Universities, Colleges, and Nonprofit Institutions (Fed Support)	Synectics/ census	https://www.nsf.gov/statistics/srvyfedsupport/; https://www.nsf.gov/statistics/srvyfedsupport/#sd;
Survey of Graduate Students and Postdoctorates in Science and Engineering (GSS)	RTI International/ census	https://www.nsf.gov/statistics/srvygradpostdoc/; https://www.nsf.gov/statistics/srvygradpostdoc/#sd;
Survey of Postdocs at Federally Funded Research and Development Centers (FFRDC PD)	RTI International/ census	https://www.nsf.gov/statistics/srvyffrdcpd/; https://www.nsf.gov/statistics/srvyffrdcpd/#sd;

continued

TABLE 6-1 Continued

Name (Acronym)	Who Collects the Data / Type	Where Documentation is Available
Survey of Science and Engineering Research Facilities (Facilities)	Westat/ census	https://www.nsf.gov/statistics/srvyfacilities/; https://www.nsf.gov/statistics/srvyfacilities/#key-survey-information&sd;
Survey of State Government Research and Development (SGRD)	Census Bureau census	https://ncsesdata.nsf.gov/sgrd/2017/; https://ncsesdata.nsf.gov/sgrd/2017/sgrd_2017_tech_notes.pdf; https://www.census.gov/programs-surveys/sgrd.html

NOTE: The National Training, Education, and Workforce Survey (NTEWS) is a new NCSES survey currently under development with plans for the survey to be performed under a negotiated Interagency Agreement with the Census Bureau.

- *InfoBriefs,* which highlight findings from recently completed surveys[1] or reports on narrowly defined topics;
- Congressionally mandated reports, such as *Women, Minorities and Persons with Disabilities in Science and Engineering* and *Science and Engineering Indicators.* For the latter report, NCSES produces the congressionally mandated *Indicators Summary Report* and cycle-specific thematic reports;
- Special recurring reports, such as *Doctorate Recipients,* based on the Survey of Earned Doctorates (SED) from U.S. universities, and *National Patterns of R&D Resources: Data Update,* which is released annually;
- Working papers;
- One-pagers, e.g., *InfoChart;*
- Infographics, e.g., *InfoBytes;* and
- Methodology reports (currently only available by request).

Due to resource limitations and time constraints, the methodology reports are given a lower priority for internal review and dissemination.

---

[1] *InfoBriefs* also provide special analysis across survey datasets or on a special topic. See the following examples: https://www.nsf.gov/statistics/2019/nsf19306/, https://www.nsf.gov/statistics/2019/nsf19300.

The data products include data tables, which are preformatted and generated by an underlying database, released with technical survey information, public-use microdata files, state profiles and academic institution profiles, interactive data tables, and restricted-use files.

More recently, NCSES has added an interactive data tool, which has a landing page where users can select a topic or type a search term. The user can also go directly to some surveys from the Build Table option. A unique feature that has been built into the search capability is one based on synonyms: instead of being limited to exact character matches, related terms can be identified using this capability, and when a user sees something of interest, he or she can select a detailed panel of relevant information. In the near future, the interactive data tool will publish the metadata application programming interface that drives this site, adding a translation layer that publishes the metadata to the Data Documentation Initiative (DDI) standards, with automated updates to Data.gov. Also, in the near future, this data tool will support an interface for downloading public-use microdata from NCSES's integrated data system.

There is also Data Explorer,[2] recently released, which facilitates exploring metadata at various levels, including measures, definitions, and variables. Finally, there is an external tool called SESTAT that provides access to public data from the Survey of Doctorate Recipients (SDR) and the National Survey of College Graduates (NSCG).

NCSES has record-level data for the following programs: Early Career Doctorates Survey (ECDS); federally funded research and development centers (FFRDCs) R&D and facilities; Higher Education Research and Development (HERD) Survey; NSCG; SDR; SED; Survey of Federal Funds for Research and Development (Fed Funds); Survey of Federal Science and Engineering Support to Universities, Colleges, and Nonprofit Institutions (Fed Support); Survey of Graduate Students and Postdoctorates in Science and Engineering (GSS); and Survey of State Government Research and Development.

Restricted survey data are available for ECDS, NSCG, SDR, and SED in the NCSES data enclave, managed by NORC at the University of Chicago on behalf of NCSES, and referred to as the Secure Data Access Facility. Researchers may apply to NCSES for access to data available within this facility. Restricted data from the Annual Business Survey (ABS) and Business Enterprise Research and Development Survey (BERD) can be accessed by researchers who apply to the Census Bureau for microdata access through federal statistical research data centers (FSRDCs). Restricted-use data for NSCG, SDR, and SED are also available through the FSRDCs.

NCSES employees who are involved in the production of surveys conducted by the Census Bureau using Title 13 or Title 26 data survey frames

---

[2]For details, see https://ncsesdata.nsf.gov/explorer.

(BERD, ABS, NSCG, SDR, and SED) can access the microdata for those studies through the Survey Sponsored Data Center (SSDC). The SSDC is a physical space within NCSES that is treated as an official Census Bureau location; the space is maintained in accordance with Census Bureau policies concerning physical and cyber-security protections for sensitive data. NCSES employees who work in the SSDC must have Special Sworn Status with the Census Bureau. Any output that is removed from the SSDC, even to be shared with other NCSES employees, must go through Census disclosure avoidance review to ensure that no information protected under Title 13 or Title 26 protected data is disclosed. The SSDCs are not designed to support research activities by NCSES employees, even those that use these same data. Census's expectation is that research projects will be undertaken in an FSRDC. With the recent expansion of virtual access for both the SSDCs and the FSRDCs, this program may be more amenable to providing the interagency transparency that is necessary to support improvements in quality and reproducibility for the NCSES data programs.

More broadly, transparency is likely to be complicated for any agencies that contract out surveys to others, including other federal agencies with their own policies, if the funding agency does not receive either survey responses or low-level aggregates needed to implement their own reproducibility and transparency-related procedures. Contracts and interagency agreements need to provide for transparency. The increasing use of remote access should make that more achievable in the future.

> **Recommendation 6.1:** Agencies that produce federal statistics should word their contracts and interagency agreements that involve planning, collection, processing, and analysis of data products so that any information adhering to the agency's transparency expectations (e.g., all processes and code related to data handling, and the resulting data collected) which is obtained by the contractor or federal agency, should be provided to the sponsoring agency unless constrained by legal or proprietary considerations.

As mentioned above, NCSES produces public-use data files, with microdata for HERD, NSCG, SDR, GSS, FFRDC R&D, and facilities. Public-use files are typically delivered from the survey contractor and reviewed and released on the NCSES Website. Additionally, the NSCG and SDR public-use data files are reproduced in house by the data team as a validation effort to check the contractor-delivered files. Several survey data files are only available under restricted access, to protect respondent confidentiality. For these data, NCSES may grant secure access to restricted-use microdata files through the Secure Data Access Facility.

NCSES also has acquired management oversight of the single-application portal known as ResearchDataGov.[3] The Census Bureau initiated the single-application portal effort and transitioned the leadership to NCSES in 2020. This portal achieves a key goal of the Foundations for Evidence-Based Policymaking Act of 2018, which is to develop a single, user-friendly online application and process for requesting data assets found in restricted data for developing evidence.

Publication Standards Utilized By NCSES

NCSES currently operates in accordance with its publication standards, as represented in their internal (not publicly available) document, "Statistical Standards for NCSES Publications." Below is the current list of standards regarding documentation of data treatments, methods, and dissemination. The four primary standards are these:

(1) Include general statements in the data sources and limitations section discussing the following:
- Sampling error (if applicable).
- Non-Sampling error
  o In general (e.g., nonresponse, errors in processing, false information provided by respondents, errors in the questionnaire, coverage or frame errors, measurement errors)
  o Any nonsampling error suspected to be associated with a particular survey.
(2) Errors that could influence the results of the analysis must be explicitly addressed (e.g., high imputation rates or editing rates for a variable used in the analysis).
(3) A description of any statistics that differs from our standard measures (means, medians, totals, ranking, percentages, ratios, percentiles [25th, 75th, etc.]) must be included within the text or as a footnote.
(4) Any methodology that differs from our standard methodology with respect to weighting procedures, or handling of non-response must be investigated and justified. A description of the methodology must be included within the text or as a footnote.

When compared with the panel's best practices tables in Chapter 7, which detail what a comprehensive approach to transparency would entail, this document leaves some things unstated regarding the archiving of record level data, details of the data collection process, where and for how long

---

[3] https://www.icpsr.umich.edu/web/pages/appfed/index.html.

the official statistics are archived, for how long input data are retained, and what metadata are used to describe these retained files. As a result, there is much left to the discretion of the survey manager, data manager, and others at NCSES involved in the collection, processing, storage, and dissemination of data.

TRANSPARENCY FOR EXTERNAL USERS OF NCSES SURVEY OUTPUT

To assess NCSES's transparency, the panel was interested in the extent to which NCSES provided information on its own Website, over and above its publication standards, to inform NCSES's external user community about various details concerning its statistical programs. Such details could include information about survey designs, the survey instruments used to collect responses, details about how to instruct the field interviewers, the extent of nonresponse and failed edits, how the survey weights were computed, the estimation methodology used, and the variability of those estimates. For this purpose, the panel decided to look at information concerning the BERD, SED, HERD, and ECDS surveys. These four surveys were chosen because of their prominence, because they represent surveys of both institutions and people, and because they are conducted by private-sector contractors as well as by the Census Bureau.

Business Enterprise Research and Development Survey (BERD)

The NCSES Web page for BERD describes the BERD survey design, which includes both simple random sampling and probability proportional to size sampling within strata, where the measure of size is based on historical research and development funding estimates. The stratification is based on research and development activity (with the most important strata being for businesses with research and development greater than $3 million), payroll, and the North American Industrial Classification System code. The sampling frame is the Business Register, maintained by the Census Bureau. BERD uses multimode data collection, a Web instrument used by 99 percent of respondents, and a paper booklet returned by mail by the remainder of the respondents.

Regarding data treatments, all data submitted are subject to several hundred automated edit checks. Approximately two-thirds of these edit checks are designed to find arithmetic errors and logically inconsistent responses, such as failed balance checks. There are also edits for outlying responses. Most failed edits are addressed using imputation; some are addressed by contacting the company. On the Website, there is also an overview of survey quality measures that is provided and there is a summary of recent changes to the survey. The degree of and treatment for unit and item nonresponse are provided, though details are left to tables in the annual reports. Coverage error is described as minimal. There is measurement error due to various differing interpretations as to what research and development activity is and due to differing accounting practices. There are also links to the survey instruments used and to the historical series of official estimates going back to 1953.

Estimation methodology is explained in the technical notes of annual detailed statistical tables and detailed in the annual methodology reports. For that reason, these BERD Web pages themselves do not detail the estimation methodology, where the weights used are a function of the sampling weights and degree of nonresponse. It would be helpful to inform users as to how the weights are calculated, but such details are also left to annual reports. Finally, there is a discussion of the comparability of the survey estimates over time given various changes to the data collection methodology, with the only substantial issues arising in 2007 and 2008.

Survey of Earned Doctorates (SED)

The NCSES Web page on SED describes this census of earned doctorate recipients. It also presents a discussion of the automated editing that is used. The only imputation for missing data that is conducted is for the month used to calculate the age of a doctorate and the time taken to complete a degree. There is also information on quality measures, which includes information on the extent of undercoverage and the degree of unit and item nonresponse, and there is some information available on the degree of measurement error. The questionnaire is also available for users to examine.

Understandably, there is no direct access to individual data due to disclosure laws, but there are two interfaces that users can use to build tables. First, for SED data from 1958 onward there is an interactive data tool that can create custom tables of the number of doctorate recipients by demographics, discipline, and institutional characteristics. Researchers can also access SED microdata through the Secure Data Access Facility and the FSRDC. The Restricted Data Analysis System (RDAS) tool, which protects confidentiality, provides access to SED data not otherwise available using the interactive data tool, which is restricted to cross-tabulations of the above-mentioned variables. However, there are plans to also access SED cross-tabulations using the RDAS tool in the near future.

Higher Education Research and Development (HERD)

The NCSES Web page on HERD contains key information about this census of educational institutions, including the questionnaire for multiple years, descriptive and methodological information such as types of edit failures and treatment for nonresponse, and quality metrics. The frame consists of all nonprofit postsecondary institutions in the United States, including Guam, Puerto Rico, and the U.S. Virgin Islands, that granted a bachelor's degree or higher in any field and which spent more than $150,000 for research and development in fiscal year 2018. Respondents may choose to respond by filling out a paper survey or by using the Web-based data collection system; 99 percent of respondents use the Web-based system. Respondents could be recontacted to address failed edits based on prior reporting patterns and unexplained missing data. Imputations for missing data are based on the previous year's data and on the data from peer institutions.

Coverage error is described as being minimal. The extent of item and unit nonresponse is also provided, and historical data are available dating back to 1972. There are some modest comparability issues over time due to changes in the census, and analysts are encouraged to contact NCSES to find out whether these issues would affect an intended analysis.

Early Career Doctorates Survey (ECDS)

The ECDS is a relatively new survey that gathers in-depth information about individuals who earned their first doctoral degree (Ph.D., MD, or equivalent) in the past 10 years and work in U.S. academic institutions or FFRDCs. The survey, described on the ECDS Web page, uses a two-stage design, the first stage being for institutions and FFRDCs, and the second covering doctorate recipients within institutions or FFRDCs. A primary stratification splits the institutions and FFRDCs into separate strata.

The selection of academic institutions is then stratified using a Carnegie classification and depending on whether the institution contains a medical school. The sampling of institutions within these strata uses "probability proportional to size" sampling based on the number of degrees historically granted per year, the number of doctorate researchers, and the number of faculty. The institutions selected then provide lists of their doctorate recipients, which are sometimes anonymized to protect privacy. Once these lists are made available, proportional allocation is used to identify the number of sampled respondents. Web survey and computer-assisted telephone interviews are offered to these respondents, the Web survey being selected more than 99 percent of the time. The questionnaire is available on the NCSES ECDS Web page. Data treatments include editing and logical and hot-deck imputation for item nonresponse. The estimates are weighted totals that account for differential sampling rates and nonresponse. Post-stratification is employed so that marginal totals agree with control totals.

For external users of the survey results, on the Web page reasons are given for possible issues leading to under- or overcoverage, and the rates of unit nonresponse are provided. Users are informed that for those interested in analyzing microdata, access to restricted data can be arranged through a licensing agreement. They are informed of the survey data quality by the release of the sampling variances and are provided with the rates of nonresponse as well. While nothing quantitative is available concerning measurement error, some cognitive testing is mentioned that was carried out to minimize this source of error.

EASE-OF-USE OF INFORMATION FOR ANALYSIS PURPOSES

The panel was also interested in ease of access to and use of the information provided, especially the ease of analysis of the official estimates themselves. Therefore, the panel asked two expert users of NCSES data for their views on the uses of the official estimates provided and associated information on issues such as the quality of the data. One expert was Anne Marie Knott, the Robert and Barbara Frick Professor in Business at the Olin School of Business, Washington University. In her presentation, Knott stated that NCSES provides important data for researchers like her who are interested in the support available for research and development. She is particularly interested in data from the Business R&D and Innovation Survey (BRDIS) (the predecessor of BERD). She initially accessed BRDIS data from a Census Bureau research data center.

One problem Knott noted when using these data in a research data center was the lack of documentation. She said it was difficult for her to determine what the precise questions were that were responded to as well as what the responses available for each question were. The documentation that was available added up to roughly 200 pages, so it was hard to find what one needed. Also, she noted, there were many versions of some variables.

In addition, Knott noted that when she was working from the publicly available estimates from the NCSES Web page, like many users of such data she was interested in examining trends over time. To carry out the desired time-series analysis, she had to merge various tables over time and then link the merged file with other covariates over time to try to relate various trends to certain explanatory variables. But such analyses were not easy, because the table numbers for different cross-tabulations change over time, and she sometimes found it difficult to determine which tables to merge. A third issue, Knott said, was that there were jumps in the data that one would assume corresponded to quantities that should be smooth over time. These jumps may have been due to changes in the survey, but documentation of such changes was difficult to find. As a result, it was hard to assess the quality of the estimates.

The second expert the panel consulted was Kimberlee Eberle-Sudre, director of Policy Research for the Association of American Universities, which comprises 63 leading research universities in the United States that collectively receive 61 percent of all federal expenditures. In her role, she accesses NCSES data at least weekly to track research funding received for use in the association's federal advocacy work, and she also uses NCSES data in the association's membership processes. Eberle-Sudre finds the data from NCSES to be essential for her work. The data have contributed substantially to the advancement of their advocacy for research at universities and colleges, she noted. As a frequent user, especially of HERD, she has become familiar with many of the nuances of both the Website and the data. She stressed that it does take some time to become comfortable with different aspects of the data tables. She believes that if NCSES modified the table presentation, it would facilitate use of the tables, especially by newer users.

Eberle-Sudre added that the cross-tabulations from the tool that replaced WebCaspar are not as user friendly as they might be. She cited three aspects of this problem: (1) one cannot save selected institutions, and as a result, each time someone wants to form a table of data on institutions, each institution must be manually selected, which can be tedious; (2) the order matters in terms of which cross-tabulation variable is selected first, second, etc.—for example, if a user chooses institution names first and then another variable, the next option may not be available since expenditure type was not selected earlier; and (3) it is difficult to locate variable descriptions, changes to variables, institutional names, and types of funding. Eberle-Sudre sees a need for a section on each program's Website that clearly lists all important changes from one survey implementation to the next and possibly another section that lists all changes to the issue under study.

While the data tables are very useful, Eberle-Sudre said, another problem is that they are designed specifically for gaining a broad descriptive view of higher education research. When a user needs to view institutional-level data or data on specific funding by federal agency, the tables are not particularly useful. She sees a need for more user involvement in the development of appropriate data tables and user interfaces.

Given the difficulties that Eberle-Sudre and other newer users have experienced, before major changes to the presentation of data are undertaken it might be useful to form a user community or network of users, which new users could turn to for mentorship and guidance. This would be particularly useful for isolated analysts of NCSES data series. (More detail is provided in Recommendations 6.1 and 6.6 below.)

PRIORITIES FOR NCSES

This panel study can be viewed as having two tasks. The first task is to examine the information that NCSES provides about its programs both internally and externally, especially the input data and official statistics and the methods used to produce those official statistics, to assess the degree to which they are currently transparent and their methods are reproducible. Further, if such an assessment demonstrates that there are some areas in which NCSES could be more open, the panel's task is to help NCSES determine what initial steps could be taken, in the most efficient manner possible, to bridge the gap between what is currently made available and what could be made available.

Second, as expressed in our statement of task, the panel has been asked to address many of the same questions for the entire federal statistical system that it has been asked to address for NCSES. As a result, in Chapter 7, as well as in previous chapters, we present a number of recommendations for ways the federal statistical agencies can more comprehensively document methods and archive their official statistics and input data. Because it is a federal statistical agency, those recommendations also apply to NCSES. However, many of those recommendations will require resources and considerable time to implement. NCSES requested that this panel also provide recommendations that are mainly relevant to NCSES that would help it begin to become more transparent in the next few years, without the addition of substantial new funding. Therefore, in the following we present recommendations designed to help bring about greater transparency for NCSES in the short term.

To begin, the panel was impressed with a number of innovative steps that NCSES has already taken to permit users to access the data underlying its cross-tabulations from its surveys to form user-requested tabulations. Such personalized use of survey data could be further enhanced by learning more about how these tools can be further refined to make them even more helpful to the public.

Recommendation 6.2: The National Center for Science and Engineering Statistics' (NCSES) information technology staff and NCSES's program staff should collaborate to develop an *ongoing program* to seek user input to improve the functionality of their Web interface, to test new analytic tools, to make it easier for users to identify documentation resources, and to facilitate appropriate access to data.

Some of the methodology reports from NCSES have not been made publicly available on their Website due to time and budgetary pressures. Examples are reports on data collection, data treatment, estimation procedures, and assessments of the quality of input data and official estimates. In order to better communicate with the public, NCSES needs to keep the public abreast of methodological improvements, investing more in the production and availability of such reports on the NCSES Web page.

Recommendation 6.3: The National Center for Science and Engineering Statistics' (NCSES) senior management should renew their emphasis on the timely production, distribution, and accessibility of methodology reports, data quality assessments, and quality profiles for each program, to ensure transparency about data quality and the information underlying NCSES's statistical programs. These studies on methodology and data quality should be a regular component of NCSES's ongoing work and be made available and accessible through the Website associated with each program.

We understand that NCSES does have a policy and process for retention of its official estimates, but we are not aware of any policy that dictates where and for how long the associated input datasets are retained. While the following recommendation pertains to NCSES, all federal statistical agencies should have a records retention policy in place, approved and monitored by the Office of Management and Budget.

Recommendation 6.4: The National Center for Science and Engineering Statistics' (NCSES) senior management should monitor and reinforce their agency's policy for archiving microdata that are the basis for the production of official statistics, as well as official statistics themselves. These policies should specify the use of data management plans, above and beyond legally required records schedules, which explicitly describe how the microdata and official statistics will meet FAIR (Findability, Accessibility, Interoperability, and Reusability) principles. The records schedule for a statistical program should always be easily accessible. The retained data and statistics should be available (via links, search capability) from the program's Website, to the extent possible consistent with confidentiality protections.

As is true for most of the federal statistical agencies, NCSES has made only modest use of shared metadata standards, in particular in its data documentation and in its exchange of official statistics and the associated methods. Given the benefits gained from sharing its official estimates with its international partners, it seems particularly beneficial for NCSES to make greater use of a metadata standard like SDMX to facilitate the transfer of its official estimates to other international statistical offices.

Recommendation 6.5: The National Center for Science and Engineering Statistics' (NCSES) information technology staff and NCSES's program staff should collaborate to standardize the inclusion of language in their contracts and interagency agreements requiring that contractors provide machine-actionable metadata and code so that NCSES can meet acceptable standards of transparency about its data products for users and other agencies and achieve consistency in the metadata used across NCSES's statistical programs. The NCSES chief statistician should monitor the implementation of this policy.

Following on our discussion in Chapter 4 of the importance of retaining paradata, NCSES should examine whether and how its programs use paradata, and based on that should examine when and for what time period paradata should be retained.

Recommendation 6.6: The National Center for Science and Engineering Statistics' (NCSES) senior management should investigate how their programs use paradata (process data, such as the number of interview contact attempts, the interim and final case dispositions, the duration of completed interviews), identify programs that would benefit from the use of paradata, identify what paradata are valuable to maintain, determine the length of time such data can be made available to researchers, and ensure that records schedules include the status of such data. While individual programs have different requirements and uses, for the purpose of transparency NCSES management should develop a policy concerning the availability and use of paradata consistent with its mission.

Given the discussion in Chapter 5 of the need to establish more regular contact with its user community, NCSES should take steps to institute greater interactions with that community to improve its perspective on how use of its data products can be facilitated.

Recommendation 6.7: Given the varying needs and expertise of different users, transparency is enhanced when all National Center for Science and Engineering Statistics (NCSES) data programs take steps to help users interact with the data used to develop official statistics. NCSES should
- Establish ongoing data user groups with contact mechanisms;
- Establish a repeated survey of users as to their current experiences in accessing and using agency data and how estimates could be presented to facilitate time series and cross-sectional analyses;
- Ensure consultations with data users prior to making changes in dissemination systems, statistical programs, and time series;
- Create a mechanism that enables members of a statistical program's user group to communicate directly with one another;

- Organize regular meetings with broad user community representation; and
- Through surveys and direct interactions with users, identify ways to improve the transparency, accessibility, and usability of NCSES estimates, data products, documentation, and dissemination systems, including the structure and navigation of the agency's Website.

7

Best Practices for Federal Statistical Agencies

BEST PRACTICES FOR DOCUMENTATION, RETENTION, RELEASE, AND ARCHIVING OF DATA

The following tables identify the information whose retention or public release, depending on the type of information, would fully support transparency in the production of a given set of official statistics. The tables therefore provide a roadmap for transparency in methods, operations, and data quality. For some types of information, this means internal retention on an agency's computer system in a permanent and internally accessible manner. For other types of information, this means the public release, such as on an agency Website or by making the information available when requested or else through release in a secure manner, for example to a federal statistical research data center.

We understand that there can be specific legal prohibitions, the need to protect proprietary information, contractual obligations, memoranda of understanding, or other constraints that could make it impossible to publicly release some information or even to keep some data internally for a period of time. On those occasions, the agencies are obligated to state publicly what has not been retained or released and why.

These tables and the accompanying recommendations are meant to be applicable to all of the principal federal statistical agencies. The Office of Management and Budget (OMB) or Interagency Council on Statistical Policy should consider monitoring how closely the principal U.S. federal statistical agencies follow these tables, acknowledging those agencies that come close to complete adherence to them.

The tables are organized as follows:

Table 7.1 Documenting Basic Elements of a Statistical Program
Table 7.2 Documenting Statistical Programs Using Survey Data
Table 7.3 Documenting Statistical Programs Using Administrative Records and/or Digital Trace Data
Table 7.4 Documenting Data Integration Issues
Table 7.5 Documenting Paradata from Statistical Programs
Table 7.6 Archiving of Data

In the tables, the leftmost column, "Information to retain," identifies the informational components. The middle and rightmost columns identify the documentation of methods and the archiving of input data and official statistics, along with associated metadata, that should be retained within a federal statistical agency (middle column), and those that should be made available to the public (right column). All of these actions support the various benefits of transparency previously discussed (see Box S-1). The metadata standards discussed and presented in Chapter 5 address a substantial percentage of the contents of these tables. For the areas that the current group of standards does not address, there is an opportunity to join the development efforts to improve the scope of the standards.

A number of lists with similar elements have been compiled, including OMB's Standards and Guidelines for Statistical Surveys, the (unpublished) Census Bureau's Statistical Quality Standards, the American Association for Public Opinion Research Code of Ethics and Practices,[1] Federal Committee on Statistical Methodology Statistical Policy Working Paper #31, and the Committee on National Statistics' Principles and Practices for a Federal Statistical Agency. The panel wanted to create this new list as an easy reference source of the elements from many of these standard documents with respect to surveys. In addition, for issues such as what to retain regarding administrative record data sources, use of digital trace data, and model-based estimates, we believe that the guidelines we submit are a reasonable start for documenting research in still developing areas.

---

[1] https://www.aapor.org/Standards-Ethics/AAPOR-Code-of-Ethics.aspx.

TABLE 7-1 Documenting Basic Elements of a Statistical Program

Information to retain or archive[a]	To be available internally, to program staff	To be available externally, to the public
The Estimation Problem: The concept or concepts this official statistical program is measuring (e.g., the percentage of U.S. households in poverty at the county level in April 2020, where poverty is defined using … and households are defined as …). This should include ✓ the precise definition of the key concepts ✓ the relevant population ✓ the levels of aggregation at which the estimates are provided ✓ the relevant time period covered by the estimates ✓ the nature of the products, e.g., tabulations, confidential, microdata, or public-use files.	Description should be updated regularly, versioned, and curated[b] for easy access. When concepts or definitions change, the data documentation for specific data products should be able to precisely connect to the version of the description that applies.	Description should be on the appropriate Web page for access by the public, updated regularly, versioned, and curated. When concepts or definitions change, the data documentation for specific data products should be able to precisely connect to the version of the description that applies. In addition, the relationship between the old and new versions should be explained for the benefit of the public.
Justification for the statistical program and input data relied on: Information required includes ✓ product sponsorship and legal authority for the data collection or program ✓ specific input datasets collected using surveys or otherwise acquired to support this estimation effort ✓ an overview of the techniques used to collect these input datasets ✓ an overview of how these datasets are used in support of the program, including weighted aggregation or any use of models, statistical or otherwise, based on these data ✓ a description of any revisions to an ongoing program, including changes to key datasets, models, methods, or procedures	Description should be updated regularly, versioned, and curated for easy access.	Information should be updated regularly, versioned, curated, and broadly publicized (e.g., on agency Website, accompaniment to estimates release, press releases, social media).

continued

TABLE 7-1 Continued

Information to retain or archive[a]	To be available internally, to program staff	To be available externally, to the public
Point of contact for information requests	Information on directing queries to the appropriate staff members should be available. It should be clear when requests for information should be directed to the agency's FOIA office and when not.	Prominent display of point-of-contact information on agency Website.

NOTES:
[a]Archiving is a more permanent and monitored process than simple saving or retention, requiring various activities to ensure continued reusability. We have not gone into each element of this series of tables to indicate whether archiving or retention is the more appropriate action; but in general, archiving would be more relevant for data and estimates and retention would be more appropriate for methodological details.
[b]One definition of curation: "The process of 'caring' for data, including organizing, describing, cleaning, enhancing and preserving data for public use. Through curation, the ICPSR provides meaningful and enduring access to data." This definition could include metadata.

Type of Data Collection

For each separately input dataset—survey (Table 7-2), administrative records, or digital trace (Table 7-3)—the following information should be saved, made public, or both. Information needed for transparency is generally the same for each data product. All documentation should be versioned and curated, and all technical reports should be made permanently available. If the reports are public facing, they should have a published DOI (Digital Object Identifier).

TABLE 7-2 Documenting Statistical Programs Using Survey Data

Information to retain or archive	To be available internally, to program staff	To be available externally, to the public
Sample design		
Target population	All details should be curated for easy access.	All releasable details should be made publicly available when estimates are released. Details that cannot be made public should be identified as such.
Sampling frame and coverage	All details should be curated for easy access.	All details should be made publicly available when estimates are released. Details that cannot be made public or are too detailed should be identified as such and made available on request.
Sampling methods, including: ✓ probability type, stratification, stages, clustering ✓ use of any optimization rules in sample design (e.g., Neyman allocation)	All details should be curated for easy access.	All details should be made publicly available when estimates are released. Details that cannot be made public or are too detailed should be identified as such and made available on request.
Sample size	All details should be curated for easy access.	All details should be made publicly available upon estimates' release.
Data collection		
✓ questionnaire employed (exact wording and skip patterns) ✓ self-administration instructions ✓ interviewer instructions ✓ languages offered ✓ self/proxy rates of a collection; identification of respondent	All details should be curated for easy access.	All details should be made publicly available upon estimates release. Details that are too voluminous should be identified as such and made available on request.

continued

TABLE 7-2 Continued

Information to retain or archive	To be available internally, to program staff	To be available externally, to the public
Modes of data collection, by percent: ✓ at the unit level ✓ including multi-mode sequence, if appropriate	All details should be curated for easy access.	All details should be made publicly available upon estimates release. Details that cannot be made public or are too voluminous should be identified as such and made available on request.
Data collection agency and dates of data collection	All details should be curated for easy access.	All details should be made publicly available upon estimates release. Details that cannot be made public or are too voluminous should be identified as such and made available upon request.
Field operation details: ✓ number of contacts per case ✓ use of incentives, including sequence if in stages ✓ final case dispositions (e.g., completed interviews, proxy interviews, imputation, refusals, noncontacts) ✓ for surveys employing adaptive design, additional information on the field operations is required[a]	All details should be curated for easy access.	All details should be made publicly available upon request. Details that cannot be made public or are too detailed should be identified as such and made available on request.
Data quality measures		
Response rate and formula employed (refer to AAPOR Standard Definitions):[b] ✓ summary statistics of case disposition by major domain	All details should be curated for easy access.	All details should be made publicly available upon estimates release. Details that cannot be made public or are too voluminous should be made available on request.
Coverage error: ✓ undercoverage, overcoverage, duplications by key domains	Detailed technical reports should be prepared or updated for each data release, versioned, and curated for easy access.	Releasable technical reports should be prepared, updated, for each data release, versioned, curated and made permanently, publicly available on agency Website with DOI.

TABLE 7-2 Continued

Information to retain or archive	To be available internally, to program staff	To be available externally, to the public
Total unit nonresponse rates, and by key domains: ✓ unit nonresponse bias reports	Technical reports should be curated for easy access.	Technical reports should be made publicly available upon release of estimates.
Item nonresponse rates by question: ✓ assessment of imputation methods ✓ analysis of item nonresponse rate by question	Details should be curated for easy access.	Releasable details should be made available upon release of estimates, as part of technical reports.
Percentage of failed edits: ✓ assessment of editing procedures	Details should be curated for easy access.	Releasable details should be made publicly available upon release of estimates, as part of technical reports.
Pretesting methods reports, including: ✓ pilot reports ✓ testing reports ✓ experiments ✓ cognitive interviews reports	Technical reports should be curated for easy access.	Technical reports should be made publicly available on agency Website.
Changes		
Changes made in survey design, survey instrument, field directions since last administration: ✓ maintain list of survey versions	A list accessible to staff should be maintained.	The list of such changes should be readily accessible by the public, as part of public technical reports, and where appropriate, on public Websites in accessible formats.
Description of data processing with commented code[c]		
Treatments for failed edits/edit specification	The code for the methodology used for treating failed edits should be retained and curated for easy access.	The general approach taken for treatments applied to address failed edits should be described and made available for the public on request. Further, the code should be commented to be readable by others and made available on request.

continued

TABLE 7-2 Continued

Information to retain or archive	To be available internally, to program staff	To be available externally, to the public
Treatment for unit nonresponse: ✓ any adjustments derived for unit nonresponse ✓ hot deck imputation, administrative records substitution	The code for the methodology used for treating unit nonresponse should be retained and curated for easy access.	The general description of the methodology used for treating unit nonresponse should be made available to the public. Further, the code should be commented to be readable by others and made available on request.
Treatments for item nonresponse	The code for the methodology used for treating item nonresponse should be retained and curated for easy access.	The general description of the methodology used for treating item nonresponse should be made available to the public.
Other post-survey adjustments: ✓ base weights ✓ undercoverage weights ✓ nonresponse weights ✓ other weight adjustments ✓ rounding, etc.	The codes for the methodology(ies) used for post-survey adjustments should be retained and curated for easy access.	The general reason for and the description of various post-survey adjustments should be made available to the public.
Transformations of variables (e.g., creation of new variables for analysis through recoding or combining multiple items)	The code(s) for the various transformations used should be retained and curated for easy access.	The description of the various transformations used and the reasons for their use should be made available for the public.
The methodology used for disclosure protection: ✓ methodology should be preserved from collection until input into final methodology used for estimation ✓ entire workflow history must be retained	The commented code to carry out disclosure protection should be retained and curated for easy access.	A high-level summary description of what is done to preserve confidentiality should be made available to the public; the disclosure avoidance methods should be released if differential privacy is employed.

TABLE 7-2 Continued

Information to retain or archive	To be available internally, to program staff	To be available externally, to the public
Methodology used to produce the official estimates	The commented code to implement the methodology used to produce the official estimates should be retained. In addition, a detailed technical description of the methodology used should be written up as a technical report suitable for publication in a technical journal and should be retained.	The summary description of the methodology used to produce the official estimates should be made available to the public. The commented code used to implement this methodology should be made available upon request.
Methods used for variance estimation	Details and commented code should be curated for easy access.	Details and commented code should be made publicly available on request.
Variability of the official estimates	A technical report providing details of the estimation of the variability of the official estimates, taking into consideration the effects of nonresponse on the input datasets used, should be retained.	A high-level report providing an outline of the estimation of the variability of the official estimates should be made available to the public.
When official estimates are the result of a model-based estimation methodology: ✓ assessment of quality of inputs used in the model	Information on what is known about the variability of the inputs should be retained.	Information on what is known about the variability of the inputs should be made available to the public.
✓ model form and related information	The form of the model, the associated parameters and how they are estimated, and assessments of the variability of the parameter estimates should be retained.	The form of the model, the associated parameter estimates and how they are estimated, and assessments of the variability of the parameter estimates should be made available to the public.

continued

TABLE 7-2 Continued

Information to retain or archive	To be available internally, to program staff	To be available externally, to the public
Relevant literature of the application of the model for this purpose	Any descriptions of the application of this type of model to analogous problems should be retained.	Any descriptions of the application of this type of model to analogous problems should be made available to the public.
Assessments of model fit, plots and tests on residuals	Any information on the fit of the model through summary statistics, residual tests, and residual plots should be retained.	Any information on the fit of the model through summary statistics, residual tests, and residual plots should be made available to the public.
Efforts to validate the model	Any efforts to apply the model to historical data, using simulated data or through use of cross-validation, should be retained.	Any efforts to apply the model to historical data, using simulated data or through use of cross-validation, should be made available to the public.
Methodology reports	Any methodology reports not included in any of the previous cells should be finalized and retained.	Any methodology reports not included in any of the previous cells should be finalized and made public on the agency Website.
Changes in methodology since last implementation	Changes in the methodology used from the previous implementation to the next should be described in a technical report and retained.	Changes in the methodology used from the previous implementation to the next should be described in a technical report and this should be made available to the public.

NOTES:
[a]See U.S. Census Bureau Statistical Quality Standards (Requirement A1-3.1) for details: https://www.census.gov/about/policies/quality/standards/standarda1.html.
[b]https://www.aapor.org/Education-Resources/For-Researchers/Poll-Survey-FAQ/Response-Rates-An-Overview.aspx.
[c]All code should default to being publicly available. Deviations (confidential parameters) should be justified, and generically identified. This applies to all rows that follow in this table.

BEST PRACTICES FOR FEDERAL STATISTICAL AGENCIES

TABLE 7-3 Documenting Statistical Programs Using Administrative Records and/or Digital Trace Data

Information to retain or archive	To be available only internally, to program staff	To be available externally, to the public
ADMINISTRATIVE RECORDS		
Target population: ✓ description of any conceptual differences between survey (if any) and administrative record source ✓ coverage of the administrative data records	All details including how they differ from the information needed in support of the official statistical product should be retained.	All details should be made permanently available to the public, versioned and curated, as part of technical reports with DOI.
Source of records and time period covered	Information should be curated for easy access.	The information should be made available to the public.
Data treatments administered	All details on how raw responses were treated prior to use should be retained.	Summary of how raw responses were treated prior to use should be made available on agency web site.
Changes made in data collection since last implementation	Any changes made to the form or the nature of the data collection since the previous implementation should be retained and any relevant technical reports on how the nature of the data elements might have changed due to changes in the external environment should be retained.	Any changes made to the form or the nature of the data collection since the previous implementation should be made available to the public, as should any relevant technical reports on how the nature of the data elements might have changed due to changes in the external environment.
Changes in the nature of the program or how people respond to the program that would impact the continuity of data from one time period to the next	All such changes should be retained.	All such changes should be made available to the public.
DIGITAL TRACE DATA		
Data disposition: ✓ source of data ✓ description of data elements ✓ conceptual link between data and information needed for statistical product, including justification for use	Descriptions of data elements and how they compare from the information needed in support of the official statistical product should be retained.	Descriptions of data elements and how they compare from the information needed in support of the official statistical product should be made available to public.

continued

TABLE 7-3 Continued

Information to retain or archive	To be available only internally, to program staff	To be available externally, to the public
How were data identified? ✓ type of data: e.g., social media, utility monitor ✓ search procedure employed (if any), e.g., Website "scraping," Internet search, cellphone records, social media sampling	The data elements and how they were found and collected should be described and retained.	Descriptions of data elements and how they compare from the information needed in support of the official statistical product should be made available to the public.
Data treatment and characteristics: ✓ cleaning and transformations, when and by whom ✓ reliability and validity of cleaned data	All data treatment techniques and results should be retained.	Descriptions of all data treatment techniques and results should be made available to the public, versioned and curated, as part of technical reports with DOI.
Changes made in data collection since last implementation	Any changes made to the form or the nature of the data collection since the previous implementation should be retained and any relevant technical reports on how the nature of the data elements might have changed due to changes in the external environment should be retained.	Any changes made to the form or the nature of the data collection since the previous implementation should be made available to the public, as should any relevant technical reports on how the nature of the data elements might have changed due to changes in the external environment.
Changes in the nature of the program or how people respond to the program that would impact the continuity of data from one time period to the next	All such changes should be retained.	All such changes should be made available to the public.
FOR BOTH ADMINISTRATIVE RECORDS AND DIGITAL TRACE DATA		
Transformations of variables (e.g., creation of new variables for analysis through recoding or combining multiple items)	The code(s) for the various transformations used should be retained and curated for easy access.	The description of the various transformations used and the reasons for their use should be made available to the public.

TABLE 7-3 Continued

Information to retain or archive	To be available only internally, to program staff	To be available externally, to the public
The methodology used for disclosure protection: ✓ should be preserved from collection until input into final methodology used for estimation ✓ entire workflow history must be retained	The commented code to carry out disclosure protection should be retained and curated for easy access.	A high-level summary description of what is done to preserve confidentiality should be made available to the public; the disclosure avoidance methods should be released if differential privacy is employed.
Methodology used to produce the official estimates	The commented code to implement the methodology used to produce the official estimates should be retained. In addition, a detailed technical description of the methodology used should be written up as a technical report suitable for publication in a technical journal and should be retained.	The summary description of the methodology used to produce the official estimates should be made available to the public. The commented code used to implement this methodology should be made available upon request.
Methods used for variance estimation	Details and commented code should be curated for easy access.	Details and commented code should be made publicly available on request.
Variability of the official estimates	A technical report providing details of the estimation of the variability of the official estimates, taking into consideration the effects of nonresponse on the input data sets used, should be retained.	A high-level report providing an outline of the estimation of the variability of the official estimates should be made available to the public.
When official estimates are the result of a model-based estimation methodology ✓ assessment of quality of inputs used in the model	Information on what is known about the variability of the inputs should be retained.	Information on what is known about the variability of the inputs should be made available to the public.

continued

TABLE 7-3 Continued

Information to retain or archive	To be available only internally, to program staff	To be available externally, to the public
✓ model form and related information	The form of the model, the associated parameters and how they are estimated, and assessments of the variability of the parameter estimates should be retained.	The form of the model, the associated parameter estimates and how they are estimated, and assessments of the variability of the parameter estimates should be made available to the public.
Relevant literature of the application of the model for this purpose	Any descriptions of the application of this type of model to analogous problems should be retained.	Any descriptions of the application of this type of model to analogous problems should be made available to the public.
Assessments of model fit, plots and tests on residuals	Any information on the fit of the model through summary statistics, residual tests, and residual plots should be retained.	Any information on the fit of the model through summary statistics, residual tests, and residual plots should be made available to the public.
Efforts to validate the model	Any efforts to apply the model to historical data, using simulated data, or through use of cross-validation, should be retained.	Any efforts to apply the model to historical data, using simulated data, or through use of cross-validation, should be made available to the public.
Methodology reports	Any methodology reports not included in any of the previous cells should be finalized and retained.	Any methodology reports not included in any of the previous cells should be finalized and made public on the agency Website.

Data Integration Issues

As mentioned in Chapter 4, Czajka and Stange (2018) have detailed a new paradigm "characterized by the use of administrative data and other forms of Big Data as alternatives to survey data… [that] necessitates new quality standards that address integrated data" (p. ix). The authors noted that many groups around the world have been studying the issue but no consensus exists for assessing the ultimate quality of an integrated data product. Rather, current research is focused on determining the quality of individual (survey, administrative, or digital trace data) components. As a result, Table 7-4 focuses on record linkage or matching, a technique being employed at most, if not all, statistical agencies and which is one of the primary integration techniques currently in use.

TABLE 7-4 Documenting Data Integration Issues

Information to retain or archive	To be available only internally, to program staff	To be available externally, to the public
Data files that were linked: ✓ identification of files ✓ description of files	A description, including the metadata, of the specific data files that were matched should be retained.	A description of the specific data files that were matched should be provided routinely as part of the technical reports or versioned data documentation.
Matching methods used: ✓ details of matching procedures	Study-specific information and technical reports should be retained.	Study-specific information and technical reports should be made available to the public.
Methods used for record linkage: ✓ methods of linking ✓ processes used to select variable sources when multiple source data sets have the same variable	The code used to carry out record linkage, along with a description of the techniques used, the variables used to match on, and a description of how the matching algorithm is implemented, including how uncertain matches are treated (sent to clerical review?) should be retained. (Note that linkage is often probabilistic, so uncertainty is built into the matching method.)	A description of the techniques used for record linkage, the variables used to match on, and a description of how the matching algorithm is implemented, including how uncertain matches are treated, should be made available to the public as part of technical reports on data quality.

continued

TABLE 7-4 Continued

Information to retain or archive	To be available only internally, to program staff	To be available externally, to the public
Evaluation of linkage success and sensitivity analysis, if used	Any information on the quality of such a match should be retained.	If available, the estimated error rates for the record linkage routine in this environment should be provided, and if not available, any information on the quality of such a match should be provided instead.

Documentation of Paradata

While paradata are typically considered for survey data, we see no reason why paradata should not be available for administrative data as well, with analogous measures (see Table 7-5). The person entering the data is analogous to the survey taker. There may be differences in availability and completeness of such measures, but they could be used for the same purposes. Also note that paradata occur during three different aspects of the data collection process: (1) paradata that result from interview contact attempts, (2) paradata that result from interview observations, and (3) paradata that result from respondent behavior.

BEST PRACTICES FOR FEDERAL STATISTICAL AGENCIES

TABLE 7-5 Documenting Paradata from Statistical Programs

Information to retain or archive	To be available only internally, to program staff	To be available externally, to the public
PARADATA FROM INTERVIEW CONTACTS		
Information on difficulty in obtaining an interview, including contact history.	Include contact history instruments employed to record results of each contact attempt. Measures based on notes or debriefs of the number of attempts needed, who the ultimate respondent was, and quality of information at each stage of data collection should be retained for as long as the information has research interest.	Information should be made available to the public on request subject to privacy considerations. For adaptive design, retention and availability of paradata are needed to justify data collection decisions that relied on them.
Information on which cases belong to which (anonymized) interviewer in order to check whether there are interviewer effects.	Information should be curated for easy access.	Information should be made available to the public on request, subject to privacy considerations.
PARADATA FROM INTERVIEW OBSERVATIONS		
For each question, computer-generated information on: ✓ the frequency of any delays, asking for assistance, visual discomfort, etc.	Measures based on notes or debriefs of interviewers on problematic questions should be retained for three years.	Information should be made available to the public on request, subject to privacy considerations.
PARADATA FROM RESPONDENT BEHAVIOR		
For each question, measurements on the respondents' degree of difficulty in responding, including ✓ click sequences for Web surveys ✓ use of various types of assistance for difficulties in responding ✓ time taken to respond ✓ degree of backtracking, etc.	Any information on measures relevant to difficulty individuals had in responding to individual survey questions should be retained.	Information should be made available to the public on request, subject to privacy considerations. Some agencies may wish to make paradata for some surveys available on a special-request basis in secure environments, like a federal statistical research data center.
Response paradata reports: ✓ response latency ✓ key stroke studies	Technical reports should be curated for easy access.	Technical reports should be made publicly available on agency Website.

Documentation for Archiving of Data[2]

All input datasets used to produce a set of official estimates and the official estimates themselves should have a records schedule and/or data management plan that indicates a permanent location for such files, how long they are to be retained, how they can be accessed, what metadata standards will be used to access them, and that these must be made public. All data products are subject to records schedule (see Table 7-6).

TABLE 7-6 Archiving of Data

Information to retain or archive	To be available only internally, to program staff	To be available externally, to the public
Archiving of treated input data and metadata (i.e., modified to account for failed edits, nonresponse, etc.)	The input datasets used to produce the official estimates (i.e., the collected data after various treatments have been applied) should be retained along with metadata that provides the record layout. (They do not need to be retained at all agencies making use of them; only the first agency producing them needs to archive.) The metadata should be machine actionable.	The input datasets used to produce the official estimates (i.e., the collected data after various treatments have been applied) should be made accessible at a secure repository, such as a federal statistical research data center, along with metadata that provide the record layout. The metadata should be machine actionable.
Archiving of untreated input data and metadata	The untreated input datasets used to produce the official estimates should also be retained along with metadata in order to support research on the treatments applied. The metadata should be machine actionable.	The untreated input datasets should be accessible at a secure repository, such as a federal statistical research data center, along with metadata that provides the record layout. The metadata should be machine actionable.

---

[2]Note that changes to the software used or to the media the data are stored on will likely make the data unreadable and, therefore, procedures also are needed to ensure that proper conversions are carried out when necessary.

TABLE 7-6 Continued

Information to retain or archive	To be available only internally, to program staff	To be available externally, to the public
Archiving of official estimates and metadata	The official estimates should be retained along with metadata that provide the record layout. The metadata should be machine actionable. The official estimates should be stored using persistent identifiers.	The official estimates should be available online for the public for a substantial time. They will be archived as per their record schedule, along with metadata that provide the record layout. The metadata should be machine actionable. The official estimates should be stored using persistent identifiers.

Recommendation 7.1: The National Center for Science and Engineering Statistics and all agencies that produce federal statistics should, to the fullest extent feasible, document their data collection methods, their data treatments, their estimation methodologies, and assessments of the quality of their official estimates, and they should archive their input datasets and their official estimates to support reproducibility and later reuse, as specified in the tables developed by the panel. To the extent possible, they should make as much of this information as possible available to their external user communities; for data treatments and estimation methodologies, they may do so through methodological overviews. They should provide reasons, such as legal or contractual constraints, for omitting items in the tables.

DEALING WITH ERRATA IN OFFICIAL STATISTICS

In discussions of the transparency of official statistics, one topic that arises is what information to provide concerning errata, which we will define to be procedural, computational, conceptual, or other kinds of errors that are discovered after release of a set of official statistics. The panel would like to distinguish between errors of different magnitudes. For example, there are errors that are relatively modest that are unlikely to change policy inferences, because the estimates have essentially the same general structure with the error retained or removed. Every series of official statistics has regular improvements of various types, and if the errors are relatively minor, it is reasonable to include with these improvements such additional "corrections" that have also been made since the last release, mentioning that the improvements also include the correction of a number of minor errors. Such improvements could be included in the release schedule for a set of official statistics, with an accompanying notification of updates of various kinds that have various sources, which could potentially include new data sources, various methodological improvements, conceptual improvements, and finally correction for small errors of various kinds.

In addition, there are on occasion (hopefully rare) more substantial errors that could result in different patterns in the estimates, which in turn could easily impact policy inferences. In such instances, we believe that it is important for transparency to call out such errors, provide their cause and nature, and release a corrected set of estimates as soon as possible and not on the above release schedule.

A VISION OF FEDERAL STATISTICS IN THE FUTURE

The panel envisions a not-too-distant future federal statistical system characterized by the implementation of more transparent methods and data, resulting in greater care in the documentation of methods, the use of uniform processes for archiving of input data and all official statistics, and the greater use of metadata standards. This will result in more sharing and reuse of input data and official statistical estimates and the methods used to produce them with the accompanying knowledge transfers within and among federal statistical agencies and with national statistical offices around the world. In this envisioned future, there will be greater interaction with the public, because today's user also wishes to make use of official statistics for nonstandard tabulations and as input to their own statistical models. As a result, agencies will have done much more in support of these alternative uses of their estimates.

Further, members of statistical programs' user communities will be more fully understood, due to increased focus on their needs (as noted at the end of Chapter 4), and they will more regularly serve as beta testers for proposed user interfaces and tools intended to provide better access to official estimates. In addition, greater use of data for research purposes will be facilitated through greater use of the federal statistical research data centers.

Internally—continuing this envisioned future—archived and documented materials will be retained in permanent Web locations and code will be fully commented and available across agencies in indicated (possibly secure) locations online. Identical machine-readable metadata standards will be used by all statistical programs, which will make sharing of methods and data easier among the statistical community. This greater sharing of data and information could extend to a variety of activities. Each program that produced official statistics would contribute to standardization and generalization by facilitating the sharing of questionnaire items, methods of coding of administrative records, applications of paradata used for common purposes (e.g., investigating nonresponse bias), or methods of combining survey and nonsurvey data.

In addition, standardized transparency will facilitate interagency collaboration on research. For instance, if one wishes to interview college graduates today, the target population is drawn from a combination of lists, but there are other surveys with similar goals, including from the National Center for Science and Engineering Statistics, the Census Bureau, and the National Center for Education Statistics, which all provide some statistics on college graduates. This could be, after a period of adjustment, addressed by sharing data across agencies.[3]

Further, because the input datasets used to produce official statistics are less likely to be survey data for many programs, and will instead use combinations of survey data, administrative data, and digital trace data, there will be the need to use sophisticated (and currently novel or unknown) models or matching techniques in producing future sets of official statistics. There is also the obvious need for more computer science expertise, both with respect to current employees and also as consultants to the agencies. This has many implications. First, the manner in which business processes operate in federal statistical agencies will be based on how other agencies have produced similar estimates, which will be straightforward given the documentation and archiving of those methods and data, respectively. Second, adding survey expertise will have a somewhat lower priority in comparison to the higher priority of providing additional expertise in statistical modeling and computer science techniques useful in documentation, archiving, and code development. There will be a need for increased interaction among the staffs of all federal statistical agencies, their user communities, and with international statistical agencies. Given the need for research into the nature and fitness for use of these novel data sources, a great deal more effort will be given to validation activities. This is also likely to require additional methodological resources.

An approach to federal statistics in which transparency and reproducibility play a larger role will be instrumental in raising the level of trust in official statistics. This would be particularly important in circumstances where normal survey operations have been disrupted.

Much of the above vision is conditional on securing additional resources, and it is also predicated on the assumption that a number of legal issues get resolved. Both are discussed below.

---

[3] See https://www2.census.gov/ces/wp/2021/CES-WP-21-19.pdf.

RESOURCE NEEDS TO PROCEED

This study was partly motivated by recent legislation, particularly the Foundations for Evidence-Based Policymaking Act, but the vision it aims for has been described in a number of earlier National Academies of Sciences, Engineering, and Medicine reports, including *Innovations in Federal Statistics: Combining Data Sources While Protecting Privacy* (2017). This vision will require additional resources, at least in the near term. Given that need, and for other reasons including various disruptions to the status quo, bringing it to fruition will require support from senior management, pilot testing, new hires, and new consulting arrangements. It is also likely that this will require new legislation, such as facilitating the collection and use of administrative data for the production of official statistics and changes to sections of the U.S. Code that prohibit specific data sharing across—or even within—federal agencies or with the public. In addition, it is important that the statistical agencies engage in the further development of statistical metadata standards, especially among all the agencies in concert and through international cooperation, such as with the United Nations Economic Commission for Europe, the Data Documentation Initiative Alliance, and the Statistical Data and Metadata Exchange.

For that reason, we have two final recommendations:

> **Recommendation 7.2:** Senior management at the agencies that produce federal statistics should provide resources and staff support to help transform their current processes to incorporate the use of data sharing and reuse through use of metadata tools and standards. This entails support for pilot projects, additional training of existing staff, enlisting of assistance from experts through support contracts, and reconfiguring of existing processes.

> **Recommendation 7.3:** Agencies that produce federal statistics, in order to implement many of the recommended initiatives in this report, should be provided with additional funds to acquire the necessary training and information technology assistance, as well as cover any increased operational costs, to modify current processes to improve documentation and archiving in support of the greater transparency of official statistics.

This report recommends that the U.S. federal statistical agencies change the way they manage metadata. All change is difficult, especially for agencies that are used to the way they conduct business. Federal budgets for the statistical agencies are, at best, flat; users want to see content additions to the products the agencies already produce, and now this report urges the agencies to take on new activities that will, at least initially, require additional funds. However, the report also recommends an incremental approach that relies on achievable goals. In this way, the report contains some advice on how to manage the change being suggested.

References

Abramitzky, R., Platt Boustan, L., Erikssen, K., Feigenbaum, J.J., and Perez, S. 2019. *Automated Linking of Historical Data*. NBER Working Paper 25825. Cambridge, MA: National Bureau of Economic Research. http://www.nber.org/papers/w25825.

Amaya, A., Biemer, P.P., and Kinyon, D. 2020. Total error in a big data world: Adapting the TSE framework to big data. *Journal of Survey Statistics and Methodology* 8(1): 89–119. doi: https://doi.org/10.1093/jssam/smz056.

American Association for Public Opinion Research. 2015. *Examples of Survey Methodological Reporting that Are Consistent with AAPOR*. https://www.aapor.org/AAPOR_Main/media/transparency-initiative/Examples_Methodological_Reporting_Revised_081015.pdf.

Arnold, T., and Kuhfeld, W.F. 2015. The StatRep System for Reproducible Research. https://support.sas.com/rnd/app/papers/statrep/statrepmanual.pdf.

Atack, J., Bateman, F., Weiss, T. 2006. *National Samples from the Census of Manufacturing: 1850, 1860, and 1870*. Inter-university Consortium for Political and Social Research [distributor]. doi: https://doi.org/10.3886/ICPSR04048.v1.

Bagley, P. 1969. *Extension of Programming Language Concepts*. Gaithersburg, MD: National Bureau of Standards.

Basker, E., Becker, R.A., Foster, L., White, T.K., and Zawicki, A. 2019. *Addressing Data Gaps: Four New Lines of Inquiry in the 2017 Economic Census*. Working Papers 19-28. Center for Economic Studies, U.S. Census Bureau.

Bates, N., Dahlhamer, J., Phipps, P., Safit, A., and Tan, L. 2010. Assessing contact history paradata quality across several federal surveys. In *Proceedings of the Survey Research Methods Section, American Statistical Association*. http://www.asasrms.org/Proceedings/y2010f.html.

Beaumont, J.F. 2020. Are probability surveys bound to disappear for the production of official statistics? *Survey Methodology, Statistics Canada*, Catalogue No. 12-001-X, Vol. 46, No. 1. http://www.statcan.gc.ca/pub/12-001-x/2020001/article/00001-eng.htm.

Becker, R.A. 2015. *Water Use and Conservation in Manufacturing: Evidence from U.S. Microdata*. Working Papers 15-16. Center for Economic Studies, U.S. Census Bureau.

Becker, R., and Grim, C. 2011. *Newly Recovered Microdata on U.S. Manufacturing Plants from the 1950s and 1960s: Some Early Glimpses.* Washington, DC: U.S. Census Bureau.

Belli, R.F., Cordova Cazar, A.L., Eck, A., and Olson, K. 2019. American Time Use Survey (ATUS): CATI Paradata, 2010. Ann Arbor, MI: Inter-university Consortium for Political and Social Research [distributor]. doi: https://doi.org/10.3886/ICPSR37318.v1.

Benguria, F., Vickers, C., and Ziebarth, N.L. 2020. Labor earnings inequality in manufacturing during the Great Depression. *Journal of Economic History* 80(2): 531–563.

Benzeval, M., Burton, J., Crossley, T.F., Fisher, P., Jäckle, A., Low, H., and Read, B. 2020. *The Idiosyncratic Impact of an Aggregate Shock: The Distributional Consequences of COVID-19.* IFS Working Paper W20/15. Institute for Social and Economic Research, University of Essex. doi: https://doi.org/10.1920/wp.ifs.2020.1520.

Bertin, A.L., Bresnahan, T.F., and Raff, D.M.G. 1996. Localized competition and the aggregation of plant-level increasing returns: Blast furnaces, 1929-1935. *Journal of Political Economy* 104(2): 241–266.

Biemer, P.P. 2010. Total survey error: Design, implementation, and evaluation. *Public Opinion Quarterly* 74(5): 817–848. doi: https://doi.org/10.1093/poq/nfq058.

Biemer, P.P., de Leeuw, E., Eckman, S., Edwards, B., Kreuter, F., Lyberg, L.E., Tucker, N.T., and West, B.T. 2017. *Total Survey Error in Practice.* New York: Wiley.

Brackstone, G. 1999. Managing data quality in a statistical agency. *Survey Methodology,* 25(2):139–149.

Bresnahan, T.F., and Raff, D.M.G. 2018. *United States Census of Manufactures, Blast Furnace Industry, 1929, 1931, 1933, 1935.* Inter-university Consortium for Political and Social Research [distributor]. doi: https://doi.org/10.3886/ICPSR37208.v1.

Callegaro, M. 2013. Paradata in Web Surveys. Pp. 259–279 in *Improving Surveys with Paradata,* edited by F. Kreuter. doi: https://doi.org/10.1002/9781118596869.ch11.

Chakrabarty, R.P., and Torres, G. 1996. American Housing Survey: A Quality Profile. *Current Housing Reports* H121/95-1. Washington, DC: U.S. Bureau of the Census.

Chun, A.Y., Heeringa, S.G., and Schouten, B. 2018. Responsive and adaptive design for survey optimization. *Journal of Official Statistics* 34(3): 581–597. doi: https://doi.org/10.2478/jos-2018-0028.

Citro, C. 2014. Principles and practices for a federal statistical agency: Why, what, and to what effect. *Statistics and Public Policy* 1(1): 51–59.

Commission on Evidence-Based Policymaking. 2017. *The Promise of Evidence-Based Policymaking: Report of the Commission on Evidence-Based Policymaking.* https://bipartisanpolicy.org/wp-content/uploads/2019/03/Full-Report-The-Promise-of-Evidence-Based-Policymaking-Report-of-the-Comission-on-Evidence-based-Policymaking.pdf.

Cunnyngham, K. 2020. *Empirical Bayes Shrinkage Estimates of State Supplemental Nutrition Assistance Program Participation Rates in Fiscal Year 2015 to Fiscal Year 2017 for All Eligible People and Working Poor People* (August). Mathematica. https://www.mathematica.org/download-media?MediaItemId={680E1B82-9AF9-4848-A857-A0048FC6A82C.

Czajka, J., and Stange, M. 2018. *Transparency in the Reporting of Quality for Integrated Data: A Review of International Standards and Guidelines.* Mathematica, April 27. https://mathematica.org/publications/transparency-in-the-reporting-of-quality-for-integrated-data-a-review-of-international-standards.

Dahlhamer, J.M., and Simile, C.M. 2009. Subunit nonresponse in the National Health Interview Survey (NHIS): An exploration using paradata. *Proceedings of the Section on Government Statistics, American Statistical Association, Joint Statistical Meeting.* http://www.asasrms.org/Proceedings/y2009/Files/302933.pdf.

Dippo, C., Fay, R., and Morganstein, D. 1984. Computing variances from complex samples with replicate weights. In *JSM Proceedings,* Survey Research Methods Section, pp. 489–494. Alexandria, VA: American Statistical Association.

European Statistical System. 2019. Quality Assurance Framework of the European Statistical System. Available at https://ec.europa.eu/eurostat/documents/64157/4392716/ESS-QAF-V2.0-final.pdf.

Fay, R. 1984. Some properties of estimates of variances based on replication methods. In *JSM Proceedings*, Survey Research Methods Section, pp. 495–500. Alexandria, VA: American Statistical Association.

Federal Committee on Statistical Methodology. 2020. *A Framework for Data Quality*. FCSM 20-04, Federal Committee on Statistical Methodology, September.

Findley, D. 2005. Some recent developments and directions in seasonal adjustment. *Journal of Official Statistics* 21(2): 343–365.

Garfinkel, S.L., and Leclerc, P. 2020. *Randomness Concerns When Deploying Differential Privacy*. Paper presented at 19th Workshop on Privacy in the Electronic Society (WPES'20), November 9, 2020 (virtual event). Available at https://arxiv.org/abs/2009.03777.

Genadek, K.R., and Alexander, J.T. 2019. *The Decennial Census Digitization and Linkage Project*. ADEP Working Paper 2019-0. Associate Directorate for Economic Programs.

Genadek, K.R., Massey, C.G., Alexander, T., Garber, T.K., and O'Hara, A. 2018. Linking the 1940 U.S. Census with modern data: Historical methods. *A Journal of Quantitative and Interdisciplinary History* 51(4): 246–257.

Gentner, D., and Kurtz, K. 2005. Learning and using relational categories. In *Categorization Inside and Outside the Laboratory*, edited by W.K. Ahn, R.L. Goldstone, B.C. Love, A.B. Markman, and P.W. Wolff. Washington, DC: APA.

Gillman, D. 2018. Using DDI and GSBPM Together at the US Bureau of Labor Statistics. UNECE Modern Stats World Workshop, Geneva, April 2018. https://unece.org/fileadmin/DAM/stats/documents/ece/ces/ge.58/2018/mtg1/USA_GILLMAN_Presentation.pdf.

Gould, W. 2011. Precision (yet again), Part I. *The Stata Blog*, June 17. https://blog.stata.com/2011/06/17/precision-yet-again-part-i.

Groves, R.M., and Lyberg, L. 2010. Total survey error: Past, present, and future. *Public Opinion Quarterly* 74(5): 849–879. doi: https://doi.org/10.1093/poq/nfq065.

Hansen, M.H., Hurvitz, W.N., and Bershad, M.A. 1961. Measurement error in censuses and surveys. *Bulletin de Institut International de Statistique* 38(2): 359–374.

Hirsch, B.T., and Schumacher, E.J. 2004. Match bias in wage gap estimates due to earnings imputation. *Journal of Labor Economics* 22(3): 689–722.

Jabine, T.B., King, K.E., and Petroni, R.J. 1990. *Survey of Income and Program Participation Quality Profile*, 2nd edition. Washington, DC: U.S. Bureau of the Census.

Kalton, G. 1998. *SIPP Quality Profile 1998*, SIPP Working Paper Number 230, 3rd edition. Washington, DC: U.S. Bureau of the Census.

Katz, L.F., and Krueger, A.B. 2019. *Understanding Trends in Alternative Work Arrangements in the United States* (January). NBER Working Paper No. w25425. Available at https://ssrn.com/abstract=3311399.

Kish, L. 1962. Studies of interviewer variance for attitudinal variables. *Journal of the American Statistical Association* 57: 92–115.

Knuth, D.E. 1992. Literate programming. *CSLI Lecture Notes*, no. 27. https://web.stanford.edu/group/cslipublications/cslipublications/site/CSIN.shtml.

Kreuter, F. 2013. Introduction. In *Improving Surveys with Paradata*, edited by F. Kreuter. Hoboken, NJ: John Wiley & Sons, Inc.

Kunz, T., and Hadler, P. 2020. *Web Paradata in Survey Research*. Mannheim, Germany: GESIS – Leibniz Institute for the Social Sciences (GESIS – Survey Guidelines). doi: 10.15465/gesis-sg_037.

Larrimore, J., Burkhauser, R.V., Feng, S., and Zayatz, L. 2008. Consistent cell means for top-coded incomes in the public use March CPS. *Journal of Economic and Social Measurement* 33(2–3): 89–128.

Lavrakas, P.J. 2008. *Encyclopedia of Survey Research Methods* (Vols. 1-0). Thousand Oaks, CA: Sage Publications. doi: https://dx.doi.org/10.4135/9781412963947.

Lenth, R.V., and Hjøsgaard, S. 2007. SASweave: Literate programming using SAS. http://homepage.divms.uiowa.edu/~rlenth/SASweave/.

Lohr, S.L., and Raghunathan, T. 2017. Combining survey data with other data sources. *Statistical Science* 32(2): 293–312.

Maitland, A., Casas-Cordero, C., and Kreuter, F. 2009. An evaluation of nonresponse bias using paradata from a health survey. *Proceedings of the Section on Government Statistics, American Statistical Association, Joint Statistical Meetings.*

Management Council of the Consultative Committee for Space Data Systems. 2012. *Reference Model for an Open Archival Information System* (OAIS). Recommended Practice CCSDS 650.0-M-2. Magenta Book. https://public.ccsds.org/pubs/650x0m2.pdf.

McCullough, B.D., and Vinod, H.D. 1999. The numerical reliability of econometric software. *Journal of Economic Literature* 37(2): 633–665. doi: https://doi.org/10.1257/jel.37.2.633.

Mulry, M.H., and Spencer, B.D. 1991. Total error in PES estimates of population. *Journal of the American Statistical Association* 86(416).

NASEM (National Academies of Sciences, Engineering, and Medicine). 2017. *Innovations in Federal Statistics: Combining Data Sources While Protecting Privacy.* Washington, DC: The National Academies Press. doi: https://doi.org/10.17226/24652.

NASEM. 2019a. *Methods to Foster Transparency and Reproducibility of Federal Statistics: Proceedings of a Workshop.* Washington, DC: The National Academies Press. doi: https://doi.org/10.17226/25305.

NASEM. 2019b. *Reproducibility and Replicability in Science.* Washington, DC: The National Academies Press. doi: https://doi.org/10.17226/25303.

NASEM. 2021. *Principles and Practices for a Federal Statistical Agency, Seventh Edition.* Washington, DC: The National Academies Press. doi: https://doi.org/10.17226/25885.

Oberski, D.L., Kirchner, A., Eckman, S., and Kreuter, F. 2017. Evaluating the quality of survey and administrative data with generalized multitrait-multimethod models. *Journal of the American Statistical Association* 112(520): 1477–1489.

Olson, K. 2013. Paradata for nonresponse adjustment. *The Annals of the American Academy of Political and Social Science* 645(1): 142–170. doi: https://doi.org/10.1177/0002716212459475.

Olson, K.M., Smyth, J., Dykema, J., Holbrook, A.L., Kreuter, F., and West, B.T. 2020. The Past, Present, and Future of Research on Interviewer Effects. pp. 1–14 in *Interviewer Effects from a Total Survey Error Perspective*, K.M. Olson, J. Smyth, J. Dykema, A.L. Holbrook, F. Kreuter, and B.T. West, editors. New York: Chapman & Hall/CRC Press.

Prasad, N.G.N., and Rao, J.N.K. 1990. The estimation of the mean squared error of small-area estimators. *Journal of the American Statistical Association* 85(409): 163–171.

Raff, D.M.G. 1988. Representative firm analysis and the character of competition: Glimpses from the Great Depression. *American Economic Review* 88(2): 57–61.

Rancourt, E. 2018. Admin-first as a statistical paradigm for Canadian official statistics: Meaning, challenges and opportunities. *Proceedings of Statistics Canada Symposium 2018—Combine to Conquer: Innovations in the Use of Multiple Sources of Data.* https://www.statcan.gc.ca/eng/conferences/symposium2018/program/03a2_rancourt-eng.pdf.

Rancourt, E. 2019. The scientific approach as a transparency enabler throughout the data life-cycle. *Statistical Journal of the IAOS* 35(4): 549–558.

Reid, G., Holmberg, A., and Zabala, F. 2017. Extending TSE to administrative data: A quality framework and case studies from Stats NZ. *Journal of Official Statistics* 33(2): 477–511. doi: http://dx.doi.org/10.1515/JOS-2017-0023.

REFERENCES

Schaeffer, N.C., Min, B.H., Purnell, T., Garbarski, D., and Dykema, J. 2018. Greeting and response: Predicting participation from the call opening. *Journal of Survey Statistics and Methodology* 6(1): 122–148. doi: https://doi.org/10.1093/jssam/smx014.

Schwanhäuser, S., Sakshaug, J.W., Kosyakova, Y., and Kreuter, F. (Eds.). 2020. Statistical identification of fraudulent interviews in surveys: Improving interviewer controls. Appendix 7 in *Interviewer Effects from a Total Survey Error Perspective*. Boca Raton: CRC Press.

SCOPE Metadata Team. 2020. *Metadata Systems for the U.S. Statistical Agencies, in Plain Language*. Prepared for Federal Committee on Statistical Methodology, July 10, 2020, Washington, DC. Available: https://nces.ed.gov/fcsm/metadata_systems.asp.

Stodden, V., Seiler, J., and Ma, Z. 2018. An empirical analysis of journal policy effectiveness for computational reproducibility. *Proceedings of the National Academy of Sciences of the United States of America* 115(11): 2584–2589. doi: https://doi.org/10.1073/pnas.1708290115.

Sundgren, B. 1973. *An Infological Approach to Data Bases*. Sweden: National Central Bureau of Statistics. https://sites.google.com/site/bosundgren/my-life/AnInfologicalApproachtoDataBases.pdf?attredirects=0.

United Nations Economic Commission for Europe. 2014. *A Suggested Framework for National Statistical Offices for Assessing the Quality of Big Data*. https://ec.europa.eu/eurostat/cros/system/files/Task%20Team%20Big%20Data%20Quality%20Framework_937_unblinded_v1.pdf.

United Nations Statistics Division. 2016. UN Statistics Quality Assurance Framework. Available at https://unstats.un.org/unsd/unsystem/Documents-March2017/UNSystem-2017-3-QAF.pdf.

U.S. Census Bureau. 2014. *American Community Survey Design and Methodology*. http://www2.census.gov/programs-surveys/acs/methodology/design_and_methodology/acs_design_methodology_report_2014.pdf.

U.S. Census Bureau. 2015. *Statistical Quality Standard F2: Providing Documentation to Support Transparency in Information Products*. Available at https://www.census.gov/about/policies/quality/standards/standardf2.html.

U.S. Census Bureau. 2019. *Current Population Survey Technical Paper*. CPS TP77. https://www2.census.gov/programs-surveys/cps/methodology/CPS-Tech-Paper-77.pdf.

U.S. Department of Commerce. 1978. *Statistical Policy Handbook*. Washington, DC. https://catalog.hathitrust.org/Record/007418500.

U.S. Office of Management and Budget (OMB). 2006. *Standards and Guidelines for Statistical Surveys*. Washington, DC. https://obamawhitehouse.archives.gov/sites/default/files/omb/inforeg/statpolicy/standards_stat_surveys.pdf.

Vardigan, M., and Whiteman, C. 2007. ICPSR meets OAIS: Applying the OAIS reference model to the social science archive context. *Archival Science* 7: 73–87. doi: https://doi.org/10.1007/s10502-006-9037-z.

Vickers, C., and Ziebarth, N.L. 2018. *United States Census of Manufactures, 1929-1935*. Inter-university Consortium for Political and Social Research [distributor]. doi: https://doi.org/10.3886/ICPSR37114.v1.

Wang, J., Kuo, T.-Y., Li, L., and Zeller, A. 2020. Assessing and restoring reproducibility of Jupyter Notebooks. In 35th IEEE/ACM International Conference on Automated Software Engineering (ASE '20), September 21–25, 2020 (virtual event), Australia. New York: ACM. doi: https://doi.org/10.1145/3324884.3416585.

West, B.T., Yan, T., Kreuter, F., Josten, M., and Schroeder, H. 2020. Examining the utility of interviewer observations on the survey response process. Chapter 8 in *Interviewer Effects from a Total Survey Error Perspective*, edited by K.M. Olson, J. Smyth, J. Dykema, A.L. Holbrook, F. Kreuter, and B.T. West, editors. New York: Taylor-Francis Group. Chapman & Hall/CRC Press.

Wilkinson, M., Dumontier, M., Aalbersberg, I., et al. 2016. The FAIR guiding principles for scientific data management and stewardship. *Scientific Data* 3:160018. doi: https://doi.org/10.1038/sdata.2016.18.

Wolter, K., and Hogan, H. 1988. *Census Coverage Evaluation: Implications of the Past and Plans for the Future*. Washington, DC: Bureau of the Census Statistical Research Division. doi: http://www.asasrms.org/Proceedings/papers/1988_019.pdf.

Zhang, L.C. 2012. Topics of statistical theory for register-based statistics and data integration. *Statistica Neerlandica* 66: 41–63. doi: https://doi.org/10.1111/j.1467-9574.2011.00508.x.

Zieschang, K.D. 1990. Sample weighting methods and estimation of totals in the Consumer Expenditure Survey. *Journal of the American Statistical Association* 85(412): 986–1001.

Appendix A

Statistical Metadata Standards—in Detail

THE UNECE FAMILY OF METADATA STANDARDS

The United Nations Economic Commission for Europe (UNECE), headquartered in Geneva, manages a number of cooperative international projects of interest to national statistical offices within UNECE and sometimes around the world. These efforts include demographic, economic, methodological, and computing-related activities. The activities sponsored by the High Level Group for the Modernization of Official Statistics (HLG-MOS) are of particular interest here. HLG-MOS oversees the development of metadata and interoperability standards developed by and for the use of national statistical offices. In the following, we discuss four of them, in this order: Generic Statistical Business Process Model, Generic Statistical Information Model, Common Statistical Production Architecture, and Common Statistical Data Architecture.

All four of these standards are equitable. The program overseen by HLG-MOS is for national statistical offices, the standards development is open to any of them, the work is conducted by consensus among the participating agencies, the work is open for inspection (transparent) on a publicly accessible wiki, and the groups of developers of the standards represent the stakeholder community in a balanced way.

Generic Statistical Business Process Model (GSBPM)

The Generic Statistical Business Process Model (GSBPM) is an outline of processes, and the names it attaches to the business processes are those a statistical agency might use to produce its data. It was produced under the auspices of the statistical program of UNECE. Representatives from national and international statistical offices throughout the world were invited to participate in the development of GSPBM under an open, consensus-driven process. As such, the GSBPM is a standard that U.S. federal statistical agencies can have confidence in and with which they can engage in its further development.

Version 4 of GSBPM was released in 2009; the current version was released in 2019 (GSBPM v5.1-UNECE Statswiki). Over this time, many countries have put it to use. It is a companion standard to the three other UNECE statistical standards described later in this Appendix: Generic Statistical Information Model (GSIM), Common Statistical Production Architecture (CSPA), and Common Statistical Data Architecture (CSDA). The relationship between GSBPM and each of these other standards is described with them.

Standards in the statistics community are possible because of the remarkable similarity of the work across national statistical offices. This is not to say that surveys are the same in every agency where they are used. Rather, an outline of the steps needed to plan, design, and conduct a survey are similar wherever they are used. So while particular questionnaires, sample designs, and variables differ across surveys, agencies, and countries, the need to design and produce questionnaires, samples, and data dictionaries persists over these divides. These similarities are the drivers behind the need to produce GSBPM and the reason it is an effective tool.

The U.S. Bureau of Labor Statistics and the U.S. Census Bureau have both adapted GSBPM to their own uses and in their own languages. GSBPM is written in English, but the international character of the group that developed the standard means some of the names for the processes described are not typically those used in the United States.

Structure of GSBPM

One way to think about a process is to compare that concept with *activity, process,* and *capability*. For the purposes of this discussion, we will define an activity as something a statistical office does, a process is how the office conducts that activity (the steps taken to accomplish a goal), and a capability is the potential for being able to carry out some activity. So, with these ideas in mind, GSBPM is not really a process model; rather, it is an activity model. And it is not really a model either; instead it is an organized list of activities. GSBPM does not tell one how to conduct the business of a statistical office; rather, it describes what needs to be done.

The GSBPM is intended to be interpreted and used flexibly. The framework is laid out in a particular way, but it is not rigid and the activities may be followed in any order. It specifies the possible steps in a statistical business life cycle. In fact, even though the activities are presented in an order typical of the work of statistical agencies, they may occur in many different orders in practice, and some steps may even be repeated in iterative applications, such as with some machine learning applications or imputation methods. Viewed as a checklist, GSBPM may be used to ensure that all necessary steps have been considered for conducting some statistical activity.

GSBPM is similar to a lattice, one that has many possible paths through it. This is what makes GSBPM generic and, therefore, widely applicable. It provides a standard view of statistical business process, yet it is neither too restrictive nor too theoretical. The GSBPM comprises three levels:

- Level 0, the statistical business life cycle;
- Level 1, the eight phases of the life cycle; and
- Level 2, the sub-activities within each phase.

GSBPM also recognizes three overarching activities that apply to all eight phases. They are as follows:

- *Quality management*—includes quality assessment and control mechanisms, recognizing the importance of evaluation and feedback;
- *Metadata management*—metadata describe each activity and describe the inputs, outputs, quality assessments, and steps used in a process to implement an activity; and
- *Data management*—includes database schemas, security, stewardship, ownership, quality, and all aspects of archiving (preservation, retention, and disposal).

The eight phases in the GSBPM are as follows: (1) *Specify needs;* (2) *Design;* (3) *Build;* (4) *Collect;* (5) *Process;* (6) *Analyze;* (7) *Disseminate;* and (8) *Evaluate.*

In the following, we describe each of these phases, but the details of the sub-activities contained within each are left out.[1] Note that, in the official GSBPM document, the term *process* rather than *activity* is used throughout. We chose to use the latter term here for clarity.

The *Specify needs* phase or high-level activity comprises six sub-activities. This phase includes all activities involving interactions with stakeholders to identify their statistical requirements (current or future). This results in proposing high-level solution options and the preparation of a business case to meet the requirements. This phase is initiated when

---

[1] We direct the interested reader to the GSBPM document on the UNECE Website at https://statswiki.unece.org/display/GSBPM/GSBPM+v5.1.

the need for new statistics is identified or feedback about current statistics is requested.

The *Design* phase also comprises six sub-activities. This phase includes development and design activities for concepts, methodologies, collection, processes, and output. This could include practical research. Included are the design features needed to define or refine the statistical products or services identified in the business case. All relevant metadata and quality assurance procedures are specified. This phase occurs at the first iteration for statistics produced on a regular basis. When improvements are identified in the *Evaluate* phase, the *Design* phase may be revisited. The *Design* phase may also make use of standards to reduce the length and cost of development and increase the ability to share, compare, and combine outputs and processes.

The *Build* phase comprises seven sub-activities. In this phase the production system is built, tested, and refined so it is ready for use. The results of the *Design* phase are configured to create the complete operational environment to run the production system. New services are built in response to gaps in the existing set of services, both from within and outside the organization. These new services are built to be broadly reusable and consistent with the business architecture of the organization. Typically, this phase occurs during the first iteration of a statistical activity.

The *Collect* phase comprises four sub-activities. In this phase all necessary information (e.g., data, metadata, and paradata) is collected or gathered by employing various collection mechanisms (e.g., acquisition, collection, extraction, transfer). Data are loaded into the appropriate environment, where further processing can occur. Validation of data formats is possible, but this does not include transformations of the data themselves. For statistics produced on a regular basis, this phase occurs in each iteration.

The *Process* phase comprises eight sub-activities. In this phase input data are processed and prepared for analysis. This includes steps to integrate, classify, check, clean, and transform input data, with the result that they can be analyzed and disseminated as statistical outputs. This phase occurs in each iteration for statistical outputs produced regularly. Data from both statistical and nonstatistical sources are included. The *Process* and *Analyze* phases can be iterative and parallel. Analysis can sometimes reveal more about the data, which might mean additional processing is necessary. Activities within the *Process* and *Analyze* phases may start before the *Collect* phase is completed. Provisional results are possible in this scenario.

The *Analyze* phase comprises five sub-activities. In this phase, statistical outputs are created and examined, statistical content prepared, and outputs determined as fit for purpose. Statistical analysts work to understand the

APPENDIX A   181

data and the statistics produced. The outputs in this phase could also be used as inputs to sub-activities in other phases. This phase occurs in every iteration for statistical outputs produced regularly.

The *Disseminate* phase comprises five sub-activities. In this phase the release of statistical products is managed. All activities undertaken to ensure release of each product, via one or more channels, are included. This phase occurs in each iteration for statistical products produced regularly.

The *Evaluate* phase comprises three sub-activities. The evaluation of a specific instance of a statistical business process is in this phase. It can take place when a process finishes and can be done on an ongoing basis. Evaluation relies on information obtained during the previous phases. Included is the evaluation of the success of a particular process, with priority on potential improvements. Evaluation should, at least in theory, occur for each iteration for regular statistical production. Determining if additional iterations are warranted is the objective. If that is decided, whether any improvements should be implemented is the next question. See Figure A-1 for a view of GSBPM in outline form.

Using the GSBPM

GSBPM is a standard recognized by the international community of national and international statistical offices. It serves as a reference model. Reference models are frameworks that consist of clearly defined concepts for promoting unambiguous communication, produced by a body of experts within some community. As such, a reference model can then be used to communicate ideas clearly among members of that same community. In the case of GSBPM, the relevant community is the national and international statistical offices throughout the world.

GSBPM is a standard without conformance criteria. It does not contain any requirements. That is reasonable, because as a reference model, GSBPM is meant for communication as guidance. It is informational in character. There are no set paths through the activities defined in GSBPM. Which activities each statistical program uses and the order in which they are used is determined by the needs of the program.

GSBPM can also be used for classification, and this represents one of the important use cases for GSBPM. Every statistical agency develops software and builds systems for processing the data surveys and other statistical programs it produces. The systems and the projects to build them can be classified according to the phases and sub-activities in GSBPM.

GSBPM is also an abstraction of the steps statistical agencies need to address to conduct surveys and other statistical programs with care. Therefore, GSBPM is very flexible. It can be used in a myriad of ways. The most direct kind of usage is to interpret the set of phases and activities literally,

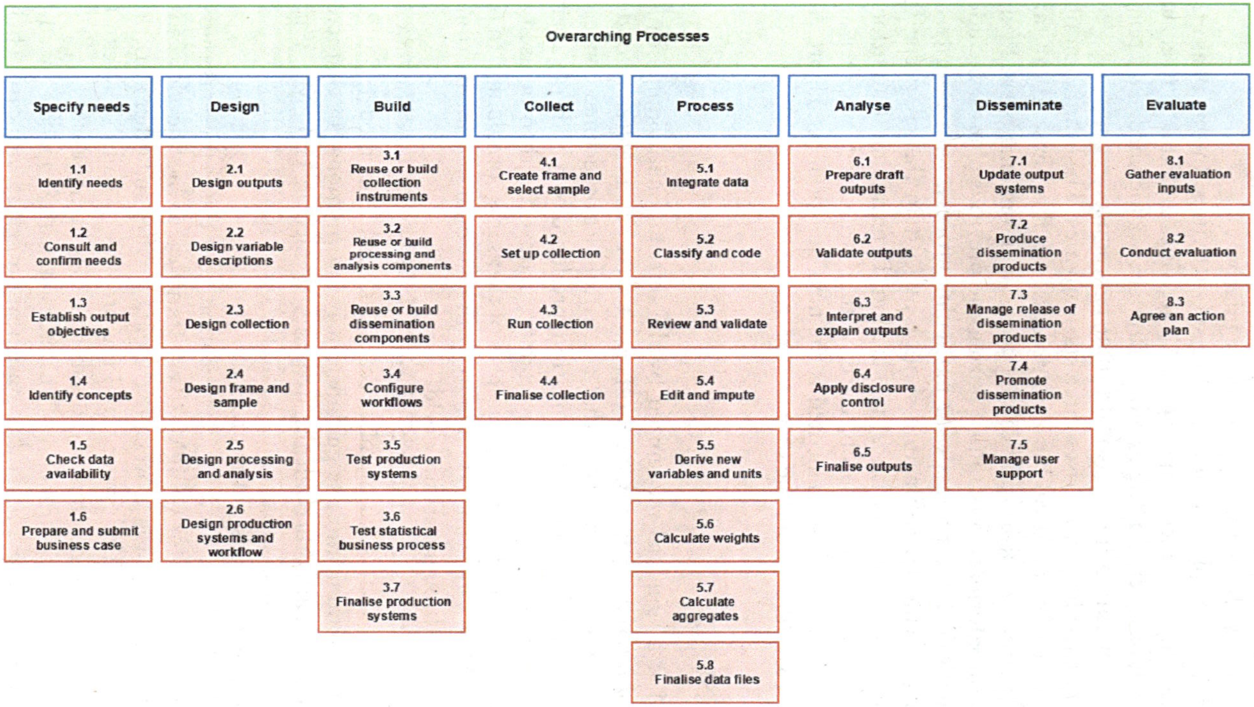

Figure A-1 GSBPM: Its processes, phases, and sub-activities.
SOURCE: United Nations Economic Commission for Europe (UNECE), on behalf of the international statistical community: https://statswiki.unece.org/display/GSBPM/Clickable+GSBPM+v5. Reproduced under Creative Commons Attribution 4.0 International License: https://creativecommons.org/licenses/by/4.0/legalcode.

Specify needs	Design	Build	Collect	Process	Analyse	Disseminate	Evaluate
1.1 Identify needs	2.1 Design outputs	3.1 Reuse or build collection instruments	4.1 Create frame and select sample	5.1 Integrate data	6.1 Prepare draft outputs	7.1 Update output systems	8.1 Gather evaluation inputs
1.2 Consult and confirm needs	2.2 Design variable descriptions	3.2 Reuse or build processing and analysis components	4.2 Set up collection	5.2 Classify and code	6.2 Validate outputs	7.2 Produce dissemination products	8.2 Conduct evaluation
1.3 Establish output objectives	2.3 Design collection	3.3 Reuse or build dissemination components	4.3 Run collection	5.3 Review and validate	6.3 Interpret and explain outputs	7.3 Manage release of dissemination products	8.3 Agree an action plan
1.4 Identify concepts	2.4 Design frame and sample	3.4 Configure workflows	4.4 Finalise collection	5.4 Edit and impute	6.4 Apply disclosure control	7.4 Promote dissemination products	
1.5 Check data availability	2.5 Design processing and analysis	3.5 Test production systems		5.5 Derive new variables and units	6.5 Finalise outputs	7.5 Manage user support	
1.6 Prepare and submit business case	2.6 Design production systems and workflow	3.6 Test statistical business process		5.6 Calculate weights			
		3.7 Finalise production systems		5.7 Calculate aggregates			
				5.8 Finalise data files			

Overarching Processes

and use the standard as a guide on how to conduct surveys and other statistical studies. However, doing this may not be useful, as the language used in GSBPM may not correspond to the terms used in each agency.

To make GSBPM useful at the U.S. Bureau of Labor Statistics (BLS), staff there reformulated the standard to meet the specific needs of the agency. The result is an outline (the model) that maps back to GSBPM but uses its own terms, phases, and sub-activities. The results appear in Figure A-2. BLS named the resulting agency standard BLSBPM. BLSBPM is in use at BLS mainly as a means to classify IT development projects and systems supporting the work of the statistical programs: censuses (e.g., QCEW), surveys (e.g., CES), and other statistical programs (e.g., CPI). Other uses for BLSBPM are planned.

The U.S. Census Bureau also developed its own version of GSBPM. This model was embedded in a larger effort called the Activity-Based Management program. This is organized as a series of top-level outlines broken into details with examples for each. The top levels are *Survey life cycle* (i.e., business process model, similar to GSBPM) and *Mission enabling and support*.

As with GSBPM, the survey life-cycle top level has eight phases associated with it. These are broken into a total of 33 sub-activities. Each of these is described with up to seven examples. The mission-enabling and support top level also is divided into eight phases, and these address work that is not directly part of the statistical life cycle. The 66 sub-activities are described with up to seven examples.

Generic Statistical Information Model (GSIM)

In the same way GSBPM depends on the similarities among the activities that agencies need to carry out to produce data and estimates, the Generic Statistical Information Model (GSIM) depends on the similarities among the kinds of objects agencies need to manage in describing their work. For example, questionnaires, sample designs, and data dictionaries are designed and produced in every traditional survey. GSIM expresses how to describe them in a uniform way.

GSIM is an internationally endorsed reference framework for statistical information developed under the auspices of UNECE, and it is an equitable standard, just as GSBPM is. This generic conceptual framework is designed, in part, to help modernize, streamline, and align the work of official statistics in and across national and international statistical offices or agencies. It is one of the building blocks for modernizing official statistics. An effect of implementing GSIM in a statistical office is the abandonment of subject-matter silos, due to the organizationwide view that GSIM promotes. This

			SAMPLING		DATA COLLECTION			ESTIMATION AND DISSEMINATION		
1 Specify needs	2 Design survey	3 Construct frame	4 Construct sample	5 Collect data	6 Review and edit collected data	7 Calculate estimates	8 Analyze estimates	9 Disseminate data	10 Archive
1.1 Analyze needs and alternatives	2.1 Design outputs	3.1 Obtain frame data	4.1 Select sample	5.1 Set up data collection	6.1 Classify and code collected data	7.1 Adjust sample weights	8.1 Review estimates and resolve anomalies	9.1 Update output systems with new estimates	
1.2 Establish measurement objectives, scope, and coverage	2.2 Identify business variables and their characteristics	3.2 Refine frame	4.1.1 Select initial sample from one or more frames	5.2 Run collection	6.2 Edit, screen, and validate collected data	7.2 Impute missing microdata	8.1.1 Review and validate estimates	8.1.2 Investigate data estimate anomalies	9.2 Produce dissemination products
1.3 Identify concepts	2.3 Design and manage data classification		4.1.3 Calculate initial sample weights	5.2.1 Request data	5.2.2 Collect data	6.3 Review and correct collected data	7.3 Derive new variables and statistical units	8.1.3 Resolve data estimate anomalies	9.3 Manage release of dissemination products
1.4 Check data availability	2.4 Design and manage data collection forms		4.2 Refine sample	5.2.3 Manage refusals and non-response	5.2.4 Monitor progress and response	6.4 Adjust collected data	7.4 Calculate aggregate weights	8.2 Apply disclosure and other publishability criteria	9.4 Market dissemination products
1.5 Determine economic / statistical methodologies	2.5 Design and manage production workflow and control		4.2.1 Make periodic adjustments to sample	4.2.2 Make sample corrections			7.5 Calculate economic estimates	8.2.1 Check for publishability based on pre-determined criteria	8.2.2 Review data for publishability	9.5 Manage data user support
						7.6 Calculate quality measures	8.3 Finalize output data		
						7.7 Calculate seasonal adjustment and factors	8.4 Produce supporting internal documentation		
						7.8 Benchmark estimates			

Figure A-2 BLS business process model.
SOURCE: Adapted from Gillman (2018).

is a fundamental precept of the strategic vision under HLG-MOS and is sanctioned by the Conference of European Statisticians.[2]

GSIM is a conceptual model containing classes and relationships that identify, describe, and relate the objects used by statistical offices to conduct statistical programs. These objects, corresponding to the classes of GSIM, are informational. They describe things statistical agencies need to manage. The data associated with the objects are metadata. This means GSIM is a statistical metadata standard.

The inputs and outputs of the processes implementing the activities of GSBPM are the classes in GSIM. This *duality* between GSIM and GSBPM is characteristic of the relationship between process and information models.

This section provides an introduction to GSIM. Some technical detail is provided, but for a full detailed technical explanation, please see the specification document and related material, available on the UNECE Website.[3]

Scope of GSIM

GSIM is written as a conceptual model, which means it makes no assumptions about other standards and technologies needed to build an implementation of the model. A conceptual model is for communicating ideas and is human readable.

The model provides a framework for describing all statistical production processes, and these are described in GSBPM. It provides names, definitions, properties, and relationships (plus related classes) for each class in the model. Each class corresponds to a useful set of objects that statistical offices should manage.

Areas such as finance, human resources, legal, and operational (building, furniture, and equipment) are not within the scope of this framework, although in some cases, when these areas touch on statistical production, those objects are described.

Structure of GSIM

When speaking of GSIM, the model presented in the standard is the main content. So, from this point on in this Appendix section, any reference to GSIM is a reference to a model. GSIM is divided into four sections, as depicted in Figure A-3. These sections are known as the top-level groups:

---

[2] See https://statswiki.unece.org/display/hlgbas.
[3] See http://www1.unece.org/stat/platform/display/metis/Generic+Statistical+Information+Model+(GSIM).

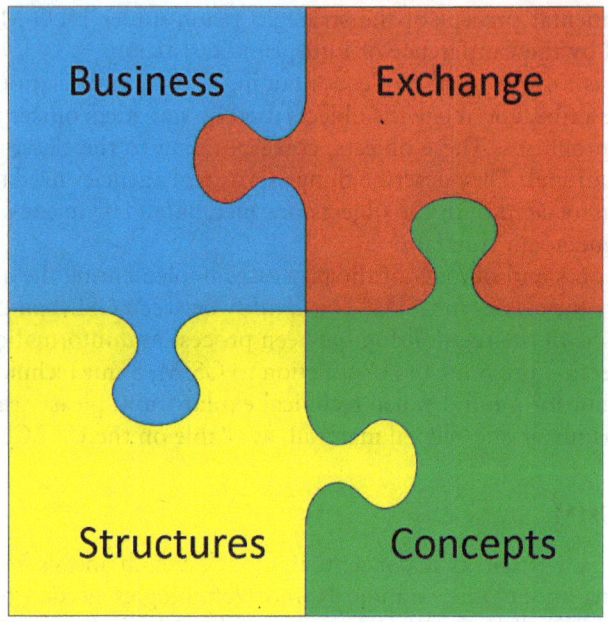

Figure A-3 GSIM top-level groups.
SOURCE: United Nations Economic Commission for Europe (UNECE), on behalf of the international statistical community: https://statswiki.unece.org/display/gsim/GSIM+v1.2+Communication+Paper (Figure 1).
Reproduced under Creative Commons Attribution 4.0 International License: https://creativecommons.org/licenses/by/4.0/legalcode.

1. The *Business* group is used to capture the planning, designs, and processes built to implement statistical programs.
2. The *Exchange* group is used to describe the data and information that are acquired and disseminated from a statistical organization via interchange mechanisms called exchange channels. Both the collection and dissemination of data can be described.
3. The *Concepts* group is used to define the basic constructs for data: concepts, variables, classifications, code lists, and their interconnections are described. Together, they convey the meaning of data.
4. The *Structures* group is used to describe how data are structured in datasets, files, and exchange channels.

Figure A-4 shows a simplified view of some of the classes identified in GSIM. It gives users examples of the kinds of objects that are in each of the four top-level groups.

APPENDIX A   187

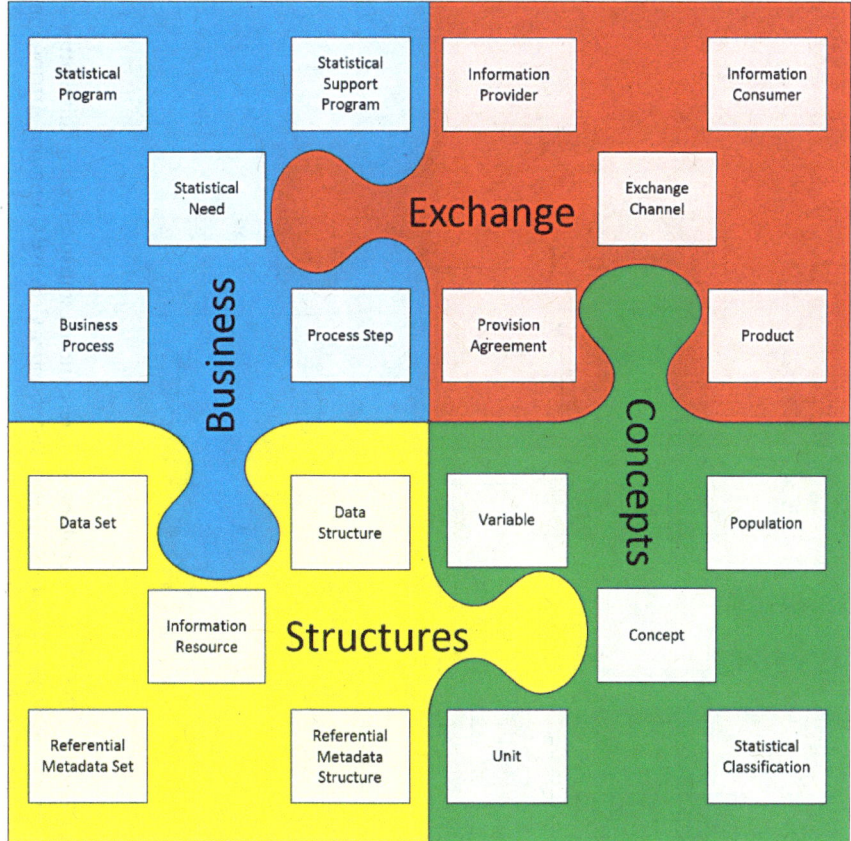

Figure A-4 Simplified view of GSIM.
SOURCE: United Nations Economic Commission for Europe (UNECE), on behalf of the international statistical community: https://statswiki.unece.org/display/gsim/GSIM+v1.2+Communication+Paper (Figure 2). Reproduced under Creative Commons Attribution 4.0 International License: https://creativecommons.org/licenses/by/4.0/legalcode.

Figure A-5 shows a slightly more technical view of GSIM. Both Figures A-4 and A-5 provide the interested reader with examples of the kinds of objects (classes) GSIM is able to manage. These correspond to the typical objects program managers need to address as part of the statistical life cycle.

Figure A-5 provides the ability to tell a story about objects associated with statistical programs that are important within a national statistical office. The story derives from the connection between objects described in GSIM and the necessary activities chosen from GSBPM. This interconnection is described next.

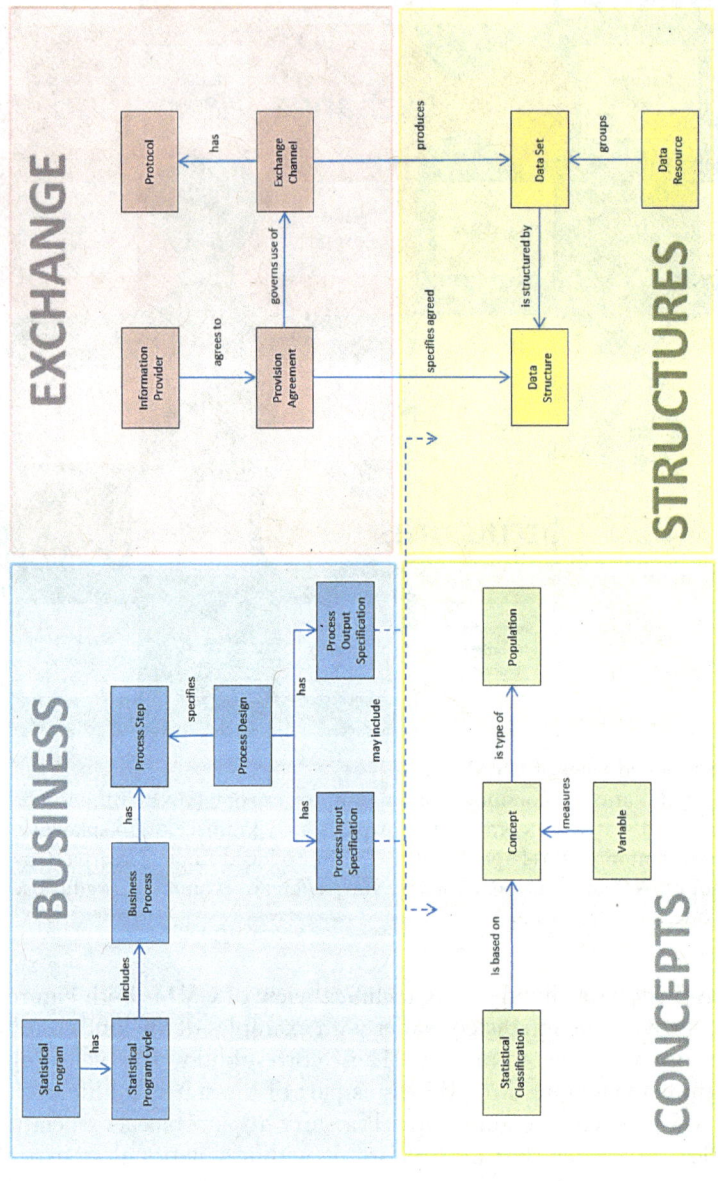

Figure A-5 Alternate but simplified view of GSIM.
SOURCE: United Nations Economic Commission for Europe (UNECE), on behalf of the international statistical community: https://statswiki.unece.org/display/gsim/GSIM+v1.2+Communication+Paper (Figure 3). (Reproduced under Creative Commons Attribution 4.0 International License: https://creativecommons.org/licenses/by/4.0/legalcode.

Interconnection with GSBPM

As briefly described above, GSIM and GSBPM are complementary or dual models. They can be used together, with GSBPM defining the activities for which GSIM describes the process implementing the activity. The inputs and outputs are described and linked to the GSIM process. This is illustrated in Figure A-6. GSIM helps describe GSBPM activities by describing the flow of information—the inputs and outputs.

Both GSIM and GSBPM are enhanced if they are used together. Greater value can be obtained, since each standard is designed for and addresses the needs of statistical agencies. Figure A-7 illustrates some specific ways GSBPM and GSIM enhance each other. Here, the levels of GSBPM are used to provide activities at ever lower detail, and these in turn define processes described in GSIM.

Restrictions on Using GSIM

In the same way that GSBPM is a reference model, so is GSIM. GSIM is a conceptual model, meant to convey to people an understanding of the information needs of a statistical office. It is not in an implementable form. Some of the details needed for an implementation were left deliberately vague in GSIM, and some parts of the model are very abstract. No specific physical representation of the model exists. Instead, GSIM provides statistical agencies with a common language and understanding related to the statistical life cycle. With GSIM, an agency has a standard way to talk about data, metadata, and all the objects the agency needs to manage.

Describing statistical information using GSIM as a common point of reference helps users (especially agencies) identify the relationship between two sets of statistical information which are represented differently from a technical perspective. This is important now that agencies are considering

Figure A-6 How GSIM and GSBPM work together.
SOURCE: United Nations Economic Commission for Europe (UNECE), on behalf of the international statistical community: https://statswiki.unece.org/display/gsim/GSIM+v1.2+Communication+Paper (Figure 4). Reproduced under Creative Commons Attribution 4.0 International License: https://creativecommons.org/licenses/by/4.0/legalcode.

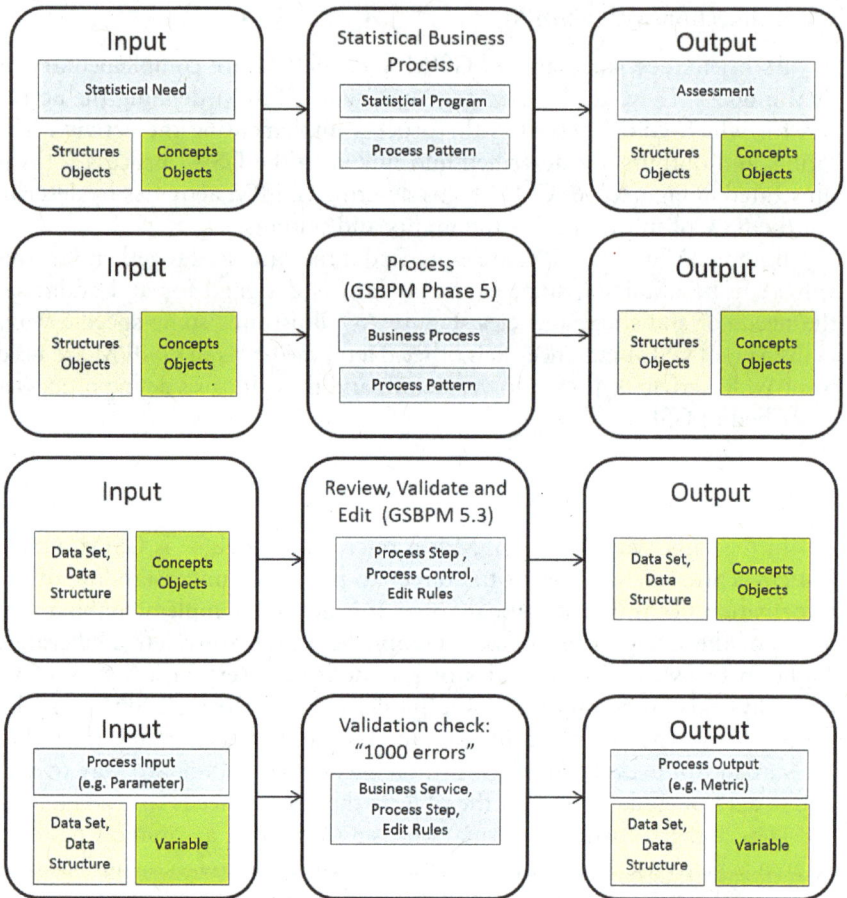

Figure A-7 GSBPM levels implemented in GSIM.
SOURCE: United Nations Economic Commission for Europe (UNECE), on behalf of the international statistical community: https://statswiki.unece.org/display/gsim/GSIM+v1.2+Communication+Paper (Figure 5). Reproduced under Creative Commons Attribution 4.0 International License: https://creativecommons.org/licenses/by/4.0/legalcode.

the needs of integrating data from multiple sources. In this case, finding the relationships between the meanings of data in different datasets is key.

Common Statistical Production Architecture (CSPA)

The Common Statistical Production Architecture (hereafter referred to as CSPA) is a reference architecture for and created by the official statistical

APPENDIX A

industry. As part of the family of UNECE standards, it has the same maintenance and governance administration as GSBPM and GSIM. Version 1 of CSPA was published and updated to version 1.5 in 2015. Version 2 is currently being drafted.

The main vision of CSPA is to enable the development of "plug and play" statistical services, which will increase the reuse of systems, thereby reducing costs and increasing the speed of implementation. This is in comparison to the traditional development of "stovepipe" or "silo" systems where the same kinds of services are developed many times throughout the statistical community to solve the same problems but in different places. An important outcome of CSPA is that the resulting service is interoperable with other CSPA services (and, to a certain extent, non-CSPA services).

CSPA consists of application and technology architectures and principles for the delivery of statistical services. One way to think of CSPA is as a service-oriented architecture (SOA) that has been specially adapted for statistical activities and processes. CSPA is not prescriptive of specific technologies or platforms.

CSPA can be combined with the GSBPM and GSIM frameworks. For example, if an agency has adopted GSBPM and it has been identified that a service is required for a GSBPM activity, then either

- the agency may develop a CSPA service, describing it in terms of the GSBPM activities it is used for, and other agencies may reuse it for those same activities, or
- the agency may identify and reuse a CSPA service already developed by another agency that matches the same GSBPM activity.

A CSPA Global Artefact Catalogue exists for such services. However, an agency may still benefit from the CSPA principles without fully implementing GSBPM or GSIM, especially to move toward a service-oriented architecture for statistical systems. The sections below illustrate the issues that CSPA is designed to address.

Accidental Architectures

When statistical agencies build systems without a standard architecture, the result is that it is very difficult for those systems to communicate with each other, and this results in "accidental architectures." It is also difficult to reuse systems across programs. For example, reusing an editing system written for one program may not work in the computing environment for another; therefore, a new system has to be developed and maintained. This new editing program may contain much of the same logic as the older one, necessitating a rewrite of the same algorithms and the implementation of another maintenance schedule, resulting in a continual waste of money and time.

These types of systems are known as silos or stove-pipe systems. If a statistical agency wishes to benefit from adopting a new technology or business process, or to align with standards, these are particularly cumbersome because there may be many duplicate systems that all have to be adapted.

The Result of Standardization within an Organization

To address the issues outlined above, an organization may adopt an enterprise architecture that can help standardize systems, processes, and interfaces within the agency. This offers the benefit of enabling the reuse of systems and services, for example the same editing software can be reused.

However, even with an enterprise architecture, this still means the statistical agency itself is operating in a silo where it cannot easily integrate a service developed by another agency. Figure A-8 (see the following subsection) shows how the Swedish systems do not fit with the Canadian systems; that is, they cannot be integrated. Indeed, most statistical agencies have very similar statistical life cycles and activities (e.g., imputation, data validation, dissemination, mapping), and very similar services have been built many times over for the same statistical processes, but those services are very hard to share and reuse. Figure A-9 illustrates this problem, in that Sweden requires a service for a particular process for dissemination that they do not have. Canada has already developed a service that proves to have exactly the functionality that Sweden needs, but it cannot be integrated by Sweden. The result is that a stove-pipe service must be created by Sweden.

To avoid this duplication of effort, what is required is a common reference architecture for both agencies. CSPA provides this so that agencies that adopt it should be able to share and integrate their services much more easily.

Using CSPA

To use CSPA as a technology architecture, an agency requires an infrastructure or platform "backbone" that allows services to be combined and run in different ways and allows the monitoring of the service runs.

CSPA has a logical information model (CSPA LIM) which is designed for developers to increase the plug-and-play compatibility of CSPA services, especially in terms of their interfaces. Full details of the CSPA LIM are on the UNECE wiki.

Adoption of CSPA also requires institutional backing, governance, and capacity building to make it a reality. For an agency to be able to efficiently

APPENDIX A 193

make a gap analysis and match available CSPA services to statistical activities, a mapping of the agency's business process model to GSBPM is required.

In order to reuse existing CSPA services, the CSPA Global Artefact Catalogue can be checked to see if an appropriate service exists.

It should be noted that CSPA is not essential for an agency to be able to reuse other agencies' systems and tools, especially if those systems support standards such as DDI or SDMX or are based on manual input. However, CSPA is designed to make it easy to integrate services that are part of tools and systems, that could be part of an automated statistical pipeline, and to avoid silos.

The CSPA architecture, usage patterns, and principles are fully described in the UNECE wiki along with a more comprehensive description of the purpose, benefits, and use cases.

Common Statistical Data Architecture (CSDA)

The Common Statistical Data Architecture (CSDA) is a reference architecture and set of guidelines for managing statistics data and metadata throughout an agency's statistical life cycle. Version 1.0 was published in 2017 and the latest version, 2.0, was published in 2018. As part of the family of UNECE standards, it has the same maintenance and governance administration as GSBPM and GSIM.

The purpose and use of the CSDA as a reference architecture is to act as a template for statistical agencies in the development of their own Enterprise Data Architectures. In turn, this will guide Solution Architects and Builders in the development of systems that will support users in doing their jobs (that is, the production of statistical products).

Data are useless without metadata, so CSDA covers data and metadata and calls them together "information." CSDA treats the physical location of the data as something that does not matter, the same principles that apply using cloud storage. CSDA stresses that statistical information should be treated as an asset. The benefits of using CSDA are

- **Independence from technology:** Statistical organization processes and systems will, eventually, become more robust to technological evolution;
- **Sustainability:** A reference architecture that is shared by the worldwide statistical community is necessary as a common vocabulary for exchanging ideas and collaboration on development and maintenance of new solutions, processes, and systems;
- **Maintainability:** Maintenance of statistical organization architectures and solutions is facilitated by the availability of a reference architecture that is shared by a larger community; and

- **Cost saving** to global optimization strategies/solutions: By referencing a shared framework, statistical organisations can better collaborate in the development, maintenance, and use of common solutions.

The CSDA standard consists of a set of key principles. Table A-1 describes each principle's rationale and implications.

The CSDA Information Model

CSDA includes an information model that consists of "Capabilities" based on the TOGAF 9 definition. A Capability is

> an ability that an organization, person, or system possesses. Capabilities are typically expressed in general and high-level terms and typically require a combination of organization, people, processes, and technology to achieve. For example, marketing, customer contact, or outbound telemarketing.[4]

The CSDA information model (see Figure A-8) comprises high-level capabilities: five core (horizontal bands) and four cross-cutting (vertical bands). These bands contain a number of lower-level capabilities (the boxes). The model and each capability are fully described in the CSDA Documentation.

The CSDA architecture is fully described in this UNECE document[5] and in the UNECE wiki, along with a more comprehensive description of the purpose and benefits and use cases.

Data Documentation Initiative (DDI)

The Data Documentation Initiative (DDI) is a family of statistical metadata standards and other work products. The work is organized under a consortium called the DDI Alliance, which is managed through a secretariat at the Inter-University Consortium for Political and Social Research (ICPSR). What follows below is a short description of each of the standards, either published or in substantial draft form.

DDI2: Codebook, version 2.5, is used to describe a social, behavioral, or economic (SBE) research study, a one-time survey or experiment, or the data each might produce. There are many applications of Codebook in

---

[4] https://pubs.opengroup.org/architecture/togaf9-doc/arch/chap03.html.
[5] https://statswiki.unece.org/display/DA/Data+Architecture+Home.

TABLE A-1 CSDA Principles: Statements, Rationales, and Implications

Principle	Statement	Rationale	Implications
1. Information is managed as an asset throughout its life cycle	• Information includes both the data and the metadata describing those data; • Information includes all objects that describe the context, content, controls and structure of data and metadata; • Information is an organizational asset that all employees have a responsibility to manage; • Information must be actively managed throughout its life cycle, from creation to disposal; • The ownership, status, quality, and security classification of information should be known at all times.	• The statistical organization has a responsibility to manage the data and metadata it acquires in accordance with relevant legalization; • Managing the information is necessary to guarantee constant quality of statistical products; • Information needs to be managed to ensure that its context and integrity are maintained over time; • As information is increasingly shared across business processes, it is important to understand the dependencies on its use.	• The statistical organization will take an enterprise approach to managing information as an asset; • Organizational policies and guidelines will be put in place to ensure that data will be managed in accordance with this principle; • All data assets will have an owner responsible for their management; • Staff will be trained to understand the value of data and their individual responsibilities; • Data quality and sensitivity will be documented where required for business processes; • Data will be protected against loss; • Data and metadata must not be kept longer than necessary, in order to protect privacy; they should be deleted at the end of their life cycles.
2. Information is accessible	• Information is discoverable and usable; • Information is available to all unless there is good reason for withholding it; • Data and metadata are accessible to humans as well as machines.	• Ready access to information leads to informed decision making and enables timely response to information needs; • Users (internal and external) can easily find information when they need it, saving time and avoiding repetition.	• The organization will foster a culture of information sharing; • Information will be open by default; • The way information is discovered and displayed will be designed with users in mind; • Systems will be designed to ensure that the minimum amount of contextual information required to understand information is captured; • Staff will create and store information in approved repositories.

continued

TABLE A-1 Continued

Principle	Statement	Rationale	Implications
3. Data are described to enable reuse	• Data must have sufficient metadata so they can be understood outside their original context; • Connections between data objects must be documented; • Restrictions to data usage must be documented.	• Data can be easily understood and used with confidence without requiring further information; • Data and their related metadata can be easily reused by other business processes, reducing the need to transform or recreate information; • The dependencies and relationships between data objects can be easily known.	• Staff will document data with reuse in mind; • Staff will consider reuse when designing systems for capturing information.
4. Information is captured and recorded at the point of creation/receipt	• Information should be captured and recorded at the earliest point in the business process to ensure that it can be used by subsequent processes; • Subsequent changes to information should be documented at the time of action.	• Information is captured and recorded at the time of creation/action so it is not lost; • The amount of information reuse is maximized by capturing it as early as possible.	• Systems will be designed to automatically capture information resulting from business processes; • Staff will need to prioritize and be given time to capture information when it is fresh in their minds.
5. Use an authoritative source	• Within a business process, there should be an authoritative source from which information should be sourced and updated; • Where practical, existing information should be reused instead of recreated or duplicated.	• Maintaining fewer sources of information is more cost effective; • Having one source of information supports discovery, reuse, and a "single version of truth."	• There will be authoritative repositories for different types of information; • Information needs will be satisfied using existing sources where possible.

| 6. Use agreed models and standards | Key information should be described using common, business-oriented models and standards agreed to by the organization. | • Having agreed models and standards will enable greater information sharing and reuse across the business process;
• Having agreed-on models and standards will enable staff to communicate using a common language. | • There will be responsibility assigned for creating and maintaining agreed-on models and standards;
• Staff will be made aware of what the approved models and standards are and how to use them;
• Agreed-on models and standards will enable external collaboration but also be fit for business purposes;
• Agreed-on models and standards will form the basis of system and process design; deviations from the standards and models will be by agreed exception only. |

SOURCE: United Nations Economic Commission for Europe (UNECE), on behalf of the international statistical community: https://statswiki.unece.org/download/attachments/314934281/CSDA%20v2.0.pdf?version=1&modificationDate=1623921475467&api=v2 (Table 1, pp 12–16). Reproduced under Creative Commons Attribution 4.0 International License: https://creativecommons.org/licenses/by/4.0/legalcode.

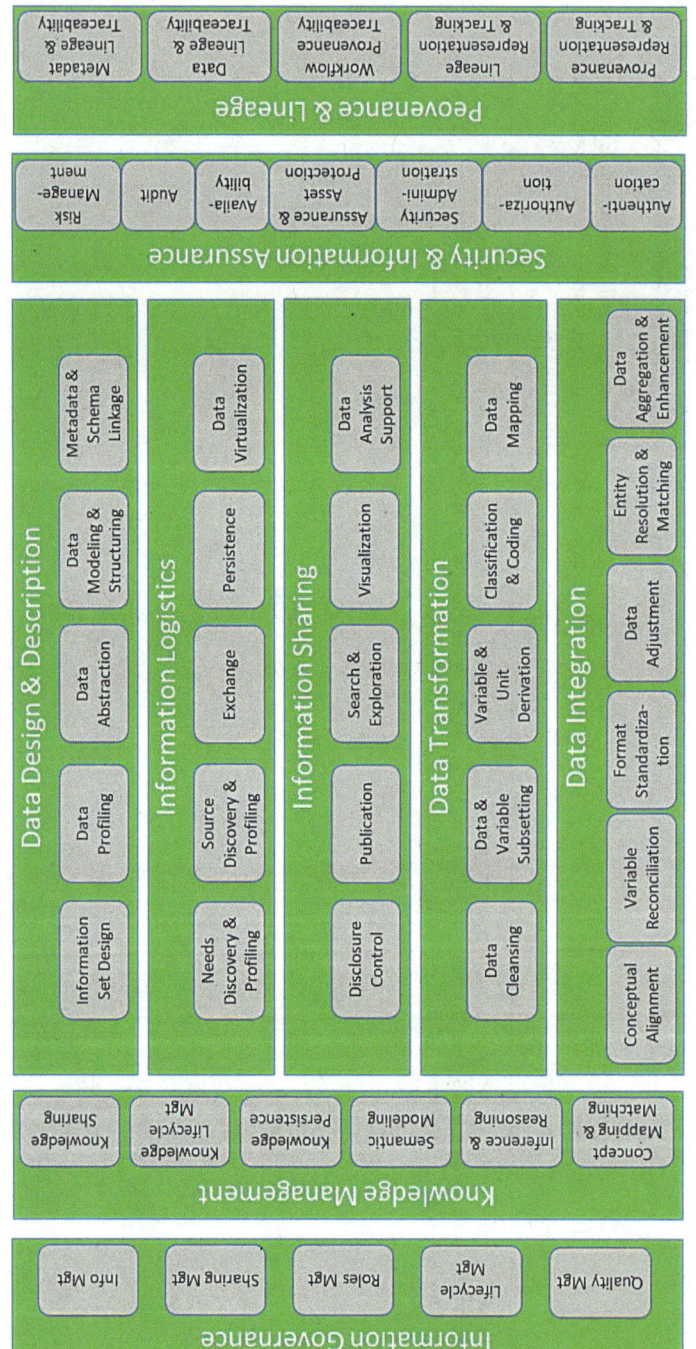

Figure A-8 Overview of capabilities and (conceptual) building blocks of CSDA.
SOURCE: United Nations Economic Commission for Europe (UNECE), on behalf of the international statistical community: https://statswiki.unece.org/download/attachments/314934281/CSDA%20v2.0.pdf?version=1&modificationDate=1623921475467&api=v2 (Figure 3, p. 29). Reproduced under Creative Commons Attribution 3.0 International License: https://creativecommons.org/licenses/by/4.0/legalcode.

use. The data archive managed by ICPSR and the International Household Survey Network, managed by the World Bank, both use Codebook as their underlying standard. However, Codebook does not support reuse of metadata, so any interconnections have to be maintained outside Codebook itself. This feature makes it easy to implement, and Codebook is often used as a first development step toward more complex metadata management systems in the statistical domain. Codebook is managed in a directly implementable form in XML.[6]

DDI3: Lifecycle, version 3.3, is used to describe the entire production cycle for statistical activities—be they censuses, surveys, or some others—conducted by national statistical offices. This capability corresponds to the work in U.S. federal statistical agencies, and the statistical life cycle is consistent with the phases of the GSBPM. Multiple surveys, each iteration of an ongoing survey, and their data can be described together, so reuse is necessary in the design of Lifecycle. Increasingly, national statistical offices around the world are turning to DDI3 for their metadata needs, including the Bureau of Labor Statistics, Statistics Canada, and the Australian Bureau of Statistics. Lifecycle is also managed as XML and is directly implementable.

DDI4: Cross-Domain Integration (DDI-CDI) was released as a substantial draft for public review and comments in April 2020. Upon resolution of all the comments, the standard is expected to be approved and released by the DDI-Alliance sometime in the latter half of 2021. The existing draft includes the ability to describe data from any source in several easily expandable logical data formats. In addition to traditional survey data (microdata and multidimensional), sources include administrative, remote sensor, Web scraping, and Internet streaming. A generalized process model is included for describing how data are processed or produced, including the provenance of data. A new datum-centered approach allows any datum to be tracked through processing or across data structures. DDI-CDI supports metadata reuse and the needs of managing data from multiple sources. The specification is managed through a UML (Unified Modeling Language) model, which allows for easy maintenance and the generation of several language representations, e.g., XML (exists), RDF[7] (in production), SQL[8] (easily generated), and others.[9]

---

[6] XML is eXtensible Markup Language, described at https://www.w3.org/XML/; for more details see http://www.ddialliance.org/Specification/DDI-Codebook/2.5/.

[7] RDF is Resource Description Framework, described at https://www.w3.org/TR/rdf-primer/.

[8] SQL is Structured Query Language, described at https://www.infoworld.com/article/3219795/what-is-sql-the-first-language-of-data-analysis.html.

[9] See https://ddi-alliance.atlassian.net/wiki/spaces/DDI4/pages/491703/Moving+Forward+Project+DDI4.

There are several other work products under DDI,[10] and these include

- DISCO—DDI-RDF Discovery vocabulary
- XKOS—eXtended Knowledge Organization System
- SDTL—Structured Data Transformation Languages
- Controlled Vocabularies.

Why Use DDI?

The rules governing standards development and participation in DDI activities, as defined for the DDI Alliance, mean that all DDI work products are equitable standards. Organizations are members of the DDI Alliance, and employees of those organizations can be designated as experts, anticipated to contribute to the work of the DDI Alliance, either technical or administrative.

Membership in the DDI Alliance allows any designated employee to participate in the technical development of DDI standards, join committees and working groups to further the Alliance or its work products, as well as vote if the member organization elects to pay its annual fee. The DDI Alliance has grown to include close to 50 members. Any organization with a material interest in the work of the DDI Alliance is encouraged to join. Several national statistical offices around the world have chosen to join the DDI Alliance, including BLS. It is expected that adoption or consideration of adopting a DDI standard is the main incentive behind joining the DDI Alliance.

The design of DDI standards is geared toward supporting several types of users. DDI encourages describing data and the programs that produce them to the fullest extent possible. Doing this supports the discovery, understanding, and usage of data (including sharing) described in this way.

DDI standards are structured, and this supports capture and maintenance of machine-readable and machine-actionable metadata.[11] This structure is a result of using the formal language for interoperability in XML, called XML-Schema.[12] Documents marked up using XML have a reusable and testable structure if the XML elements are defined using XML-Schema. In this case, there is a formal validation procedure to make sure documents follow all the rules. DDI standards take advantage of this feature.

---

[10]The interested reader can explore these specifications on the DDI Website at https://ddialliance.org/Specification/RDF for DISCO and XKOS, at https://ddialliance.org/products/developing-products-of-the-alliance for SDTL and again for DISCO, and at https://ddialliance.org/controlled-vocabularies for controlled vocabularies.

[11]See https://www.icpsr.umich.edu/web/pages/about/continuous-capture.html for a project led by George Alter to operationalize continuous capture of metadata.

[12]XML-Schema is described at https://www.w3schools.com/xml/schema_intro.asp.

This means DDI standards are implementable immediately. Anyone with the tools to use XML-Schema and XML can create, manipulate, validate, maintain, and use documents structured in XML. This differentiates the DDI standards from GSIM. GSIM is a conceptual model expressed in UML. It is not written in an immediately implementable form. In addition, GSIM is missing some details in areas that some of the DDI standards provide, such as the detailed description of multidimensional data in DDI-CDI. However, the combination of DDI-Lifecycle and DDI-CDI incorporates GSIM.

As mentioned above, the DDI standards support several kinds of users: librarians, archivists, researchers, software developers, standards developers, managers, survey methodologists, and subject-matter experts. As the potential application areas for DDI standards increase, more types of users will need to be included, thus increasing the kinds of organizations with a material interest in the work of the DDI Alliance.

DDI Data Life Cycle

The data life cycle in use in the DDI community (see Figure A-9) appears very similar to the set of phases as laid out in GSBPM. This is purposeful, because DDI-Lifecycle is intended to incorporate the survey life cycle in use in national statistical offices. However, there are a few differences in focus.

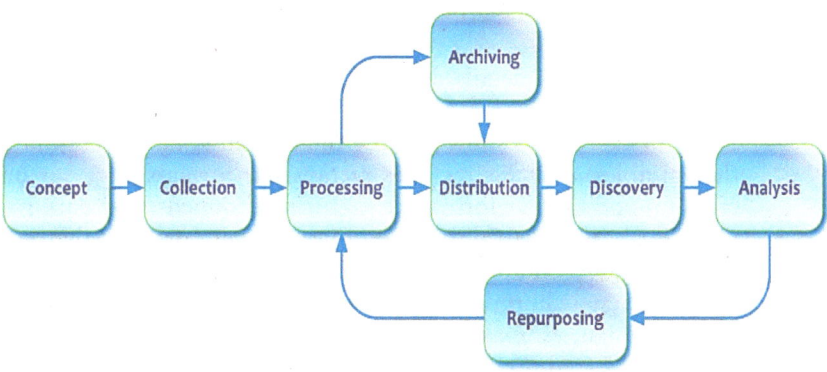

Figure A-9 Data life cycle as conceived in DDI Data Lifecycle.
SOURCE: United Nations Economic Commission for Europe (UNECE), on behalf of the international statistical community, https://ddi-lifecycle-documentation.readthedocs.io/en/latest/User%20Guide/Introduction.html. Reproduced under Creative Commons Attribution 4.0 International License: https://creativecommons.org/licenses/by/4.0/legalcode.

The main differences are around Archiving and what happens to data after they are disseminated (Distribution in DDI). Since the DDI standards were originally developed to help data archives, this is not a surprise, and this phase is provided to address these needs. A bigger difference is in supporting Discovery and Analysis after achieving Distribution. This is outside the scope of GSBPM, but it is an important part of the data life cycle for librarians, archivists, and researchers.

DDI Standards

DDI-Codebook. The DDI-Codebook (Codebook) was the first work product produced under DDI. It is being used throughout the world. As a result, the standard continues to be maintained with corrections added as they are found. The latest version of the standard is labeled 2.5, the "2" signifying Codebook.

As explained earlier, Codebook does not support reuse. The focus is on a class of objects called *Study*, described one at a time. Think of a Study as a research project, based on a one-time survey or experiment. What follows, then, is a short discussion of the objects needed by Codebook to describe that Study.

The main kinds of metadata needed to describe a Study are

- *Contact*—information about subject-matter experts (e.g., the researcher conducting a Study);
- *Study information*—basic and high-level descriptions of a Study;
- *Variables*—details describing the variables needed to describe the data produced;
- *File information*—details describing data files (there can be more than one);
- *File format*—details showing how the data in a file correspond to the variables, and this includes rectangular formats (e.g., for microdata) and tabular formats (e.g., for multidimensional data);
- *Questions*—the questions from a questionnaire or form that lead to the variables, if applicable; and
- *Methodology*—short descriptions of methodological issues, such as sample design, estimation procedures, analytical procedures (e.g., imputation), and processing (e.g., editing and classification).

There are other details as well. The interested reader is encouraged to visit the DDI-Codebook pages on the DDI Alliance Website under "Specifications" (https://ddialliance.org/explore-documentation).

DDI-Lifecycle. DDI-Lifecycle (Lifecycle) resulted from more demanding requirements uncovered through the use of Codebook, especially as the needs of national statistical offices and support for the statistical survey life cycle were recognized. Support for the phases of GSBPM and the reuse requirements for describing ongoing surveys (and not just one-time studies) were major factors.

Lifecycle has many additions that differentiate it from Codebook. The idea of a Study persists, but now Studies fit into larger groupings and contain smaller groupings that provide a very flexible means for linking surveys and other statistical programs to the relevant metadata needed in each case. Some metadata, such as statistical classifications, are not developed with a single survey in mind. They are sharable. Lifecycle supports this idea, and this is a prime example of the idea of reuse.

This means that a complex statistical program, such as the Consumer Expenditure Surveys (CE) series at BLS, is describable. CE is conducted through two ongoing data collections, the quarterly Interview Survey and the weekly Diary. The Interview Survey is conducted monthly on a rotating sample, and the Diary is collected weekly, with each sampled household providing two consecutive weeks of data. Interview households are in the sample for one year, so each household is interviewed four times.

Data are processed through four post-collection editing, imputation, and allocation phases, with several activities included under each. Finalized data are ultimately sent to the Consumer Price Index program every month and combined in quarterly estimates to produce yearly estimates every six months. Yearly microdata files are also produced, one for public use and one with confidential information attached for restricted use.

Changes to the surveys occur in the odd years, though some very small corrections may be made at any time. So, metadata describing CE need to cross surveys, concepts, designs, time, processing, and datasets. Codebook is not up to this task, but Lifecycle was designed to handle this kind of complexity.

Lifecycle versions are designated with a "3"; the current release is 3.3. As with Codebook, Lifecycle was built and is maintained as a series of XML-Schemas. Therefore, it can be implemented directly, just as Codebook can. Commercial, open-source, and shareable software produced in various offices and universities exist to help federal agencies implement Lifecycle. The implementation describing CE at BLS took advantage of this, and the system there implements Lifecycle 3.3.

There are many areas of increased detail, reuse, and management supported by Lifecycle that are not available through Codebook. Some of these are the following:

- *Variable cascade* – The ability to describe variables in four sharable levels. These levels are
  - Concept—the concept defining a variable, such as *marital status*, which of course can be shared;
  - Conceptual variable—the additional concepts associated with a variable, e.g., allowed categories (single, married, widowed, divorced for marital status) and the universe, e.g., adults;
  - Represented variable—the additional codes or representations the categories, numeric ranges, or textual constraints needed, including the data type as intended; and
  - Variable—the use of a variable in a dataset or application, such as an SAS® dataset; the codes representing missing data are added here.
- *Design considerations*—Questionnaire design, sampling plan, and weighting are all included, each providing substantial detail. For instance, it is possible to describe each stage of a multistage sample.
- *Questionnaires*—The wording of questions, response choices, and question flow are all part of the description. This makes it possible to use Lifecycle as the framework for building a complex questionnaire before it is sent to developers to build an instrument, for example using Blaise® or some other system. This has the potential for substantially shortening the development time of questionnaires and their instruments.
- *Groups*—Kinds of metadata can be grouped for sharing. For instance, if the same set of variables is used to describe the data in datasets representing each of the major statistical packages, it is very inefficient to link each variable to each, especially if this does not change over time as well. By assigning the relevant variables to a group, one associates the variables just by linking the group.
- The ability to describe the flow of a questionnaire is very similar to describing the steps in the processing of data. So, processing steps are describable using this feature. This means Lifecycle can be used to describe all the phases in GSBPM.

DDI–Cross Domain Integration (DDI-CDI). DDI–Cross Domain Integration (DDI-CDI) is the latest in the family of DDI standards, though at this writing it is still in draft form. The final release of the standard is expected in late 2021. Readers of this report should assume the current draft is substantially correct and complete and that the draft available prior to its release will be quite similar to that version that is released. Versions of DDI-CDI will be designated with "4," so the first release will be 4.0.

DDI-CDI represents some significant changes to the management and scope of the DDI standards. First, DDI-CDI is built as a UML[13] model. This means the standard must be "serialized" into an implementable framework if it is to be used immediately. The current draft release contains such an XML serialization, based again on XML-Schema. There are plans to build serializations in RDF, SQL, and others as time permits. But DDI-CDI is immediately implementable, just as Codebook and Lifecycle are. At this time, there is no external software that supports DDI-CDI such as there is for Codebook and Lifecycle.

The decision to use UML was based on the realization that it is easier to maintain and modify a UML model than a collection of XML-Schemas. UML has many technical advantages for describing metadata, the most important being the ease with which UML represents relationships among classes of objects. These relationships are what make reuse possible. For XML-Schemas, relationships across schemas are not a natural part of that standard. XML-Schemas express rules for hierarchies, so UML is a more natural fit. The reason XML-Schema was chosen for Codebook and Lifecycle is that XML can be used directly to build a system.

The reasons behind the development of DDI-CDI are somewhat varied. At first, the idea was to express Codebook and Lifecycle in UML and add new capabilities that neither currently has into one manageable system. The work ended up as an attempt to redesign Codebook and Lifecycle from scratch. This turned out to be too ambitious.

At the same time, members of the development team for DDI-CDI began consulting with people outside the DDI community to make sure DDI-CDI is compatible with other standards and the needs of other domains. The need for a cross-domain specification became apparent during this work. Serendipitously, efforts in the statistical community showed the need for combining data from multiple sources, thus the idea for a new standard within the DDI family, this time DDI-CDI.

Several World Wide Web Consortium[14] (W3C) standards were tested against DDI-CDI for compatibility. The standards DCAT (for a data catalog) and PROV (for dataset provenance) are especially popular and useful. DDI-CDI conforms to DCAT and PROV. PROV was especially important since the provenance of datasets outside the statistical domain is so important for understanding how to use them.

DDI-CDI is intended to address the ever-increasing need for integration of data from multiple sources. Codebook and Lifecycle are designed for describing SBE data. DDI-CDI is more generic, intended to describe data from many domains. Interesting and useful data are produced by

---

[13]UML is Unified Modeling Language. It is described at https://www.uml.org/.
[14]https://www.w3.org/.

many organizations and for many purposes. DDI-CDI is able to describe datasets, organized in a variety of ways, with any provenance, independent of subject, and independent of technology.

DDI-CDI borrows heavily from Lifecycle, but it expands the descriptive capability in new and detailed ways. Here is a list of the several innovations in DDI-CDI:

- Expanded and more detailed description of variables. The variable cascade is more carefully modeled. It relies on linking to concepts more than before, and the management of subject-matter categories (substantive) and missing categories (sentinel) clearly separates them;
- Understanding data, universes, populations, categories, and variables as concepts is fundamental. Now, a datum can be followed across datasets and through processing—the datum-centered approach;
- Expanded process model for describing how data are produced and their provenance, from any domain;
- Ability to describe ordered and unordered collections or groupings of objects;
- Expanded ways data can be structured in files, including rectangular (the typical way in statistics), long (for event history data or a data warehouse), and key-value pairings (for Web scraping, sensor data, satellite imagery, etc.);
- New description of multidimensional data, providing an integrated approach to describing time series and n-cubes; and
- Borrowed and expanded idea from GSIM to assign roles to variables (measures, auxiliary attributes, identifiers, dimensions) for analytic purposes. The roles are not fixed, the same variable may be assigned a different role in another context, and variables are used to convey to another user which data are under analysis.

Statistical Data and Metadata eXchange (SDMX)

The main purpose of the Statistical Data and Metadata eXchange (SDMX) standard is to format multidimensional data and metadata into a framework for automated exchange among organizations. The accompanying metadata make the data semantically interoperable, so SDMX supports full interoperability of multidimensional statistical data. See https://sdmx.org/ for more details.

The standard was developed by seven international statistical offices and banks: Bank of International Settlements (BIS), European Central Bank (ECB), Eurostat (the statistical office of the European Union), International Monetary Fund (IMF), Organisation for Economic Co-operation and Development (OECD), United Nations Statistics Division (UNSD), and World Bank. It undergoes periodic updates, and a new version will be released soon.

SDMX was approved as an international statistical standard in 2008. The United Nations Statistical Commission at its 39th session "recognized and supported SDMX as the preferred standard for the exchange and sharing of data and metadata, requested that the sponsors continue their work on this initiative and encouraged further SDMX implementations by national and international statistical organizations."[15] SDMX was also approved as a technical specification by ISO/TC154 in ISO TS 17369 in 2013. This ISO approval means users can have confidence that SDMX products are reliable and of good quality.

From the technical perspective, SDMX is an integrated solution consisting of three main elements:

- technical standards (including the Information Model),
- statistical-content-oriented guidelines, and
- IT architecture and tools.

These three elements can be implemented in a stepwise approach.

An SDMX exchange package includes a Data Structure Definition (DSD), and its construction is the elemental aspect of using the standard. A DSD consists of a description of the measure, the dimensions associated with it that describe some multidimensional data, and attributes that describe additional information about the data. Dimensions and attributes may be coded (using an associated code list) or not. Each code list representing a dimension has an identifier (or, in the parlance of the Web, a Uniform Resource Identifier, or URI) that allows interoperable exchange.

SDMX incorporated the idea of a registry to handle the DSDs, code lists, and other resources needed to make exchanges work in a seamless (interoperable) way between partners. The scope of a registry can be global or less so (parochial). The global registry holds resources that may be used within exchanges anywhere in the world. More parochial registries, say one for the U.S. federal statistical agencies or one for just the Census Bureau, BEA, BLS, or any other agency, are for the data and metadata each agency exchanges. SDMX includes a standard for registries, so different software may provide the function of a registry but they should all be compatible with other SDMX registries and tools.

Some dimensions are common across all subject domains, and these are called cross-domain code lists. These are maintained among the content-oriented guidelines. Some DSDs (known as global DSDs) are used by many

---

[15] https://sdmx.org/?sdmx_news=un-statistical-commission-sdmx-is-preferred-standard.

organizations, because there are internationally agreed ways to represent some multidimensional data. Both the cross-domain code lists and global DSDs are maintained in the global registry. The global registry identifier for each of its entries is its URI.

Since the initial publication of SDMX in 2004, many tools have been developed to assist implementers in using the standard. Some are devoted to the use of the registries, both global and parochial, and all are being improved and fine-tuned regularly. Most tools are free and open source. The number of global implementations is growing, covering many domains in statistics, and the existence of the tools has helped considerably in the expansion of the systems. The current developments are geared toward making SDMX easier to use and implement, and they broaden the features and data types of SDMX.

In this context, statistical agencies may be confronted with the question of considering SDMX as a solution for harmonizing and automating their multidimensional data and metadata exchanges with international organizations or within their own organization. For instance, the U.S. agencies providing national indicators to the IMF Statistical Data Bulletin Board do so through the U.S. Department of the Treasury via SDMX. Several years ago, BLS, BEA, and the Board of Governors of the Federal Reserve built a pilot system for exchanging some U.S. national indicators with the ECB, although that system was never fully developed. Even agencies that have not yet adopted SDMX may reconsider their decision, taking into account the significant progress made over the years.

Typical Use Cases and Scenarios for SDMX

SDMX can support various use cases and implementation scenarios, as described below. To do so, a set of common tools, processes, terminologies, and methodologies to facilitate the exchange of information between producers and consumers has been produced. We describe two important use cases here.

The most typical use case for SDMX is data and metadata reporting. This is the case where an agency exchanges some data with another statistical agency or organization. The circumstances for why the exchange happens vary, of course, but the fundamentals of an exchange are the same. The infrastructure used by SDMX to achieve this objective contains a network of registries that make DSDs and code lists available for use. Also, a set of reporting formats and IT tools to assist implementers are included. Full implementation of such infrastructure results in direct machine-to-machine exchange.

Data discovery and visualization is another important use case. Here, the objective is to make statistical data and metadata findable and accessible by external users. Under this scenario, queries based on the SDMX information model drive Website presentation of data and metadata. Based on user selections, the application retrieves the requested information (data and metadata) from SDMX structural repositories, and transforms it into tables, graphs, charts, etc.

Also, SDMX can be used to consistently model data and metadata (e.g., an agency's internal data) in order to improve harmonization and reduce the metadata maintenance burden. This can help reduce duplicate data storage, improve metadata quality, and enable linkages between datasets.

Benefits of SDMX

Due to the similar nature of the statistical activities across all national and international statistical organizations, many face similar challenges. Solutions to these challenges vary, but implementing standards in statistical offices has advantages, as Chapter 5 addresses. SDMX provides some benefits beyond those of other standards, and we discuss these here. These areas are

- Harmonization of statistical data and metadata,
- Cost reduction to end users who build applications for accessing data,
- Data quality improvement through faster validation, and
- Reduction in the reporting burden for data providers.

These benefits are described through a discussion around some tools and resources that are part of the SDMX system of tools and the content-oriented guidelines.

Harmonization. At its core, harmonization requires that the meaning of data be presented in a way that can be compared. SDMX is designed to do this in a natural and automated way. Dimensions are a fundamental part of multidimensional data, and these are managed as code lists in SDMX. Code lists are made available for inspection and comparison through the network of registries.

Codes and terms play the same terminological role in code lists. They represent the concepts (categories) in the code lists, and they are used to convey those meanings just as words do in natural language. And as with natural language, some terms or codes are spelled the same but have different meanings (homographs), while some terms or codes are spelled differently but have the same meaning (synonyms). Examples are many, but as one example: in gender and marital status code lists, the string 'm' can represent *male* (as a gender) and *married* (as a marital status). This is an example of a homograph. Similarly, different systems might use their own representations for a gender code list, for example one might use 'm' for male and the other might use '1' instead, even though the meaning is the same; this is an example of synonyms.

Harmonization is the process of cleaning up these ambiguities, and SDMX, through the network of registries, directly supports this activity. This is achieved in part by making differences visible, creating the pressure to remove differences that are gratuitous (having no substantive meaning). As a code list in the network of registries becomes more global, data using that code list as a dimension are more harmonized or standardized.

The SDMX glossary provides a common resource to understand the terms used in SDMX, and this helps provide an unambiguous interpretation of the standard. For any standard, the number of varieties it allows, based on multiple interpretations, is a measure of how hard it is to use. The glossary reduces this inconsistency, and this makes each implementation more like all the others. Importantly, each glossary term is coded so that it may be consistently used in data structures, mappings, and machine-actionable processing.

The advantages of harmonizing and standardizing statistical data and metadata content and structure are numerous:

- Implementers and users speak the same language.
- Reusing existing material saves time and resources.
- Reuse is facilitated by the existence of SDMX registries.
- There is less mapping and data processing.
- Numerous existing SDMX objects are available for defining data and metadata.
- Tools based on a commonly agreed format have a wider audience.

SDMX improves interpretability because it harmonizes structural metadata (the identifiers and descriptors of data, such as table columns and stubs) and their terminology (the SDMX Glossary). SDMX, therefore, contributes to the development of a global statistical language (along the lines of the models developed under the auspices of UNECE: GSBPM, GSIM, CSPA, and CSDA).

SDMX can also be said to improve coherence through the use of cross-domain concepts, shared code lists, harmonized statistical guidelines, and the extensive reuse of SDMX objects across domains and agencies.

But harmonization comes at a cost. On the technical side, unifying disparate systems is usually a major project that needs proper planning and allocation of resources. On the substantive side, developing common classifications and code lists requires an ongoing organizationwide effort that individual departments may see as interfering with their work. These challenges are inherent in any harmonization effort.

APPENDIX A

Reducing Costs. SDMX can reduce IT development and maintenance costs, mainly through its open-source approach, as many SDMX tools exist. Other open-source tools are available: Statistical Information System-Collaboration Community.Stat Suite, SDMX Reference Infrastructure, and Fusion Registry. It is strongly recommended to consider these tools before developing a new platform. SDMX has always taken seriously the idea that different organizations will implement at their own speed and with their own objectives. The result of this is the "toolkit" approach: SDMX offers many different tools, but they need not all be adopted or used together. Differentiated implementation strategies are thus possible, making the standard accessible to countries or entities of varying capacity levels.

A large community offering to share expertise around the standard exists as well. Furthermore, sharing resources worldwide is better than working nationally, as it promotes cross-fertilization of ideas and practices. This open-source approach thus means no licensing costs, a shared toolbox, and the sharing of development burden among the international community.

SDMX has a strong user base, sponsors yearly user conferences, and has active user groups, and the standard has broad support. All these factors help to reduce the rather steep learning curve that using SDMX entails. Further, as a reminder to the reader, SDMX does not have as broad a scope as some of the other standards discussed in this Appendix. More than just SDMX is required to capture the metadata for the entire statistical life cycle.

Data validation. SDMX natively provides structural validation of data and metadata. This means that a data message can be checked to be sure it matches the structure described by the DSD and dataflow and does not contain invalid dimensions or codes.

To enable data content validation, the SDMX community developed the Validation and Transformation Language (VTL) standard. The purpose of VTL is to allow a formal and standard definition of algorithms to validate statistical data and calculate derived data.

The language is designed for users who may not have information technology (IT) skills (such as statisticians) who should be able to define calculations and validations independently, without the intervention of IT personnel. It is based on a "user" perspective and a "user" information model (IM) and not on possible IT perspectives. As much as possible, the language is able to manipulate statistical data at an abstract/conceptual level, independently of the IT representation used to store or exchange the data.

The language is intuitive and friendly (users should be able to define and understand validations and transformations as easily as possible), so the syntax is

- designed according to mathematics;
- expressed in English to be shareable in all countries;
- as simple, intuitive, and self-explanatory as possible;
- based on common mathematical expressions; and
- designed with minimal redundancies.

The language is oriented to statistics. For example, it contains

- operators for data validations and edit;
- operators for aggregation, even according to hierarchies; and
- operators for dimensional processing (e.g., projection, filter).

At a later stage, it contains

- operators for time series processing (e.g., moving average, seasonal adjustment), and
- operators for statistics (e.g., aggregation, mean, percentiles, variance, indexes).

The language is designed to be applied not only to SDMX but also to other standards. To varying degrees, it can work with GSIM and DDI. In particular, it can operate on all the various data structures that GSIM and DDI describe.

Reducing the reporting burden. SDMX is designed to significantly reduce the effort and resources required by reporting agencies in a typical reporting framework. It does this in the following ways:

- SDMX enables the "pulling" of data and metadata from a single cross-domain exit point, which is the standard SDMX Web interface. A reporting agency that implements SDMX may replace multiple reporting systems with the SDMX Web interface. Data collection becomes more practical for data collectors, as they can query the data as they want rather than downloading a full snapshot of a dataset, and the query format is the same across SDMX-implementing agencies.
- Efficiencies can be gained by avoiding round trips of checking and fixing data between reporter and collector agencies, because SDMX enables the validation of data structure and coding before and after the data transmission (by the reporter and collector). The validation can be easily automated, and existing tools may be used by all agencies.
- Traditional exchange frameworks often rely on multiple, bespoke exchange agreements because of the different formats required by

each agency and the different times collectors may require the data. SDMX can simplify and reduce the exchange agreements required by harmonizing the data structures and formats. For example, an exchange agreement can reference a global DSD, such as for National Accounts. Reporting calendars and frequencies may be simplified, as the reporter can publish the data as soon as they are ready and the collecting agencies can "pull" the data from the agency's SDMX Web interface as and when required.

Comparing VTL with DDI SDTL. The Structured Data Transformation Language (SDTL) under DDI is not the same as VTL. SDTL does not support validation. VTL is designed to be executable, whereas SDTL is for documentation. SDTL is readable through a set of structured JSON (JavaScript Object Notation) tags, but VTL must be parsed based on syntax rules. Finally, SDTL includes a schema and software to translate into a natural language description.

GENERAL METADATA STANDARDS

There are a number of highly adopted metadata standards (many from the library and digital curation communities). These commonly used standards may be invoked with other standards, or they may be worth using as part of an application profile.

Dublin Core Metadata Element Set (DCMES)

The Dublin Core Metadata Element Set (DCMES) is a "vocabulary of fifteen properties for use in resource description."[16] Though it was created initially with the goal of supporting the description of born-digital resources such as digital images, digital documents, e-books, and so on, it also can be used to describe physical objects (e.g., paintings, books, DVDs, etc.). It can also be used to describe datasets, albeit with some limitations (described further below).

Development of Dublin Core began in 1995 at an OCLC/NCSA workshop in Dublin, Ohio.[17] The "core" refers to the broad nature of the standard: the terms are intended to be generic enough to be applied to a wide range of resources. The 15 elements are contributor, coverage, creator, date, description, format, identifier, language, publisher, relation, rights, source, subject, title, and type. Descriptions of each element and notes on

---

[16]https://www.dublincore.org/specifications/dublin-core/dces/.
[17]See the following page for additional details on this: http://www.dlib.org/dlib/July95/07weibel.html.

how to use them can be found on the Dublin Core Metadata Initiative (DCMI) Website.

When people talk about Dublin Core, they are most often referring to the 15-element Metadata Element Set. However, there are several additional vocabularies created and maintained by the Dublin Core Metadata Initiative that are important to be aware of as well. Two to be aware of are these:

- DCMI Metadata Terms [DCMI-TERMS]: This is the set of all the terms/elements defined by the DCMI, including the Metadata Element Set, as well as "several dozen properties, classes, datatypes, and vocabulary encoding schemes."
- DCMI Type Vocabulary [DCMI-TYPE]: This is a controlled vocabulary for describing the "nature or genre"[18] of a resource, e.g., "Physical Object," "Dataset," etc.

Dublin Core is meant to describe resources at the *file level*. Key to DC's design is the "1-to-1 principle." That is, for every resource being described, there should be one Dublin Core metadata record. This is meant to facilitate disambiguation between variations of the same resource, e.g., between a thumbnail image and the main version of a file. A common example considers multiple versions of the Mona Lisa: the original painting, a high-resolution .tiff photograph of the painting; and a smaller .jpg thumbnail. For each of these resources, there should be a corresponding Dublin Core metadata record. While this may seem straightforward, this can potentially lead to situations in which the creator of the tiff of the Mona Lisa is not listed as Leonardo Da Vinci—but rather, the name of the photographer who took the photo. Thus, it is important to clearly articulate a workflow for accurately cataloging different versions of resources and linking records accordingly.

Dublin Core is one of the most commonly adopted metadata standards in academic digital repositories and digital libraries. The generic focus of the element set makes it broadly applicable to most collections, particularly collections of documents, books, musical works, artistic works, and so on. Because it can describe both physical and digital objects, it is popular with libraries, archives, and museums. However, the generic nature of the elements means that Dublin Core is not necessarily useful for repositories seeking to serve a very specific community or use. This standard does not have any official standing with any national or international statistical agencies, but it is worth noting that it has been ratified as an ISO standard.

With regard to describing datasets, Dublin Core can be used to create very general metadata, but it is not designed to provide much variable-level

---

[18]https://www.dublincore.org/specifications/dublin-core/dcmi-terms/#section-7.

documentation, or even to provide much description of methodological considerations. Again, Dublin Core is meant to describe resources at the file level, which means a Dublin Core metadata record for a dataset would be intended to describe the dataset as a whole—not specific variables or observations within the dataset. Consider again the 15 terms in this element set:

> contributor, coverage, creator, date, description, format, identifier, language, publisher, relation, rights, source, subject, title, type

Many of these can be adapted to be used to describe statistical datasets, but the fit may be somewhat awkward. For instance, "coverage" can be used to describe the geospatial and temporal range of a dataset. But this is truly meant to encompass a range, such as a geometric polygon on a map, rather than specific geolocalities. The "description" field is broad enough that it could be used to describe all the variables within a dataset, but this would not necessarily be machine readable without specific adaptations. This general, file-level approach means that Dublin Core could potentially be used to describe disparate types of statistical dataset, such as administrative records, quality reports, or various types of text-based data, but only at the aggregate (dataset) level.

Dublin Core is often used as part of a metadata "application profile," in which multiple standards are combined for a bespoke use, and/or are applied in a unique way. It is likely that Dublin Core could be used in combination with another standard to provide the interoperability that it affords in combination with more domain-specific applications. Dublin Core also has numerous published crosswalks to other standards, notably from the Getty Center, an institution focused on visual art and cultural heritage.[19] However, these crosswalks typically focus on other standards intended for libraries, archives, and museums, rather than ones such as the others covered in this Appendix.

Software support for Dublin Core is focused in the areas in which it is heavily used. Several digital library and digital asset management platforms have built-in support for Dublin Core, including Fedora and Omeka. Fedora is "a robust, modular, open-source repository system for the management and dissemination of digital content. It is especially suited for digital libraries and archives, both for access and preservation."[20] Similarly, Omeka is an open-source Web publishing platform for the organization and display of archival, library, and digital scholarship collections.[21] However, Dublin Core is not produced, handled, or otherwise supported by any

---

[19] https://www.getty.edu/about/.
[20] https://duraspace.org/fedora/about/.
[21] https://omeka.org/.

statistical software. Its metadata can be serialized into HTML, XHTML, XML, and XML/RDF syntax, with XML being a common choice.

One considerable upside of Dublin Core is ease of use. It is among the easier standards to implement and customize, which is one reason for its popularity. Every Dublin Core record refers to one, and only one, object and consists of a number of attribute-value pairs. Implementing Dublin Core can become somewhat more involved when combining it with other standards in an application profile, as mentioned above, but when used on its own it is fairly straightforward and minimally complex. Because Dublin Core does not require any proprietary software or platform to use, it would also be minimally costly to any institution that wished to begin using it. While Dublin Core can only be effectively applied to statistical data at the dataset level, it has the advantage of applying equally well and with equal ease to any type of dataset.

Overall, Dublin Core is an unusual standard to include in a report about statistical data. It is included here not only because it is an important metadata standard, broadly speaking, but also because it is a simple, general standard that may be worth using in tandem with another standard covered here to record metadata at the dataset level. In this role, Dublin Core could be used in order to capture and document potentially important information at the aggregate level and facilitate the organization and retrieval of statistical datasets. While it is possible to find discussion of Dublin Core and its use in many places on the Web, particularly among library and information science communities, the primary resource and reference for Dublin Core is the DCMI website, specifically the terms list.[22]

PREMIS (PREservation Metadata: Implementation Strategies)

Developed by the Library of Congress, PREMIS (PREservation Metadata: Implementation Strategies) is "the international standard for metadata to support the preservation of digital objects and ensure their long-term usability."[23] Simply put, PREMIS specializes in expressing preservation metadata about digital objects—"the information a repository uses to support the digital preservation process."[24] The PREMIS data dictionary defines a core set of concepts needed by repositories to perform preservation actions on their digital collections. This includes information about formats, implementations, hardware and media, agents, rights, and provenance. PREMIS is rarely used as a stand-alone standard but, rather,

---

[22] https://www.dublincore.org/specifications/dublin-core/dcmi-terms/.
[23] See Reference: https://www.loc.gov/standards/premis/.
[24] From "Understanding PREMIS," https://www.loc.gov/standards/premis/understanding-premis.pdf.

is typically used in combination with other standards that provide more descriptive metadata. It therefore could likely be used in conjunction with the metadata standards described below.

METS (Metadata Encoding and Transmission Standard)

METS (the Metadata Encoding and Transmission Standard) was also developed by the Library of Congress. It is a "standard for encoding descriptive, administrative, and structural metadata regarding objects within a digital library"[25] and is often used as a "wrapper" for other metadata from other schemas. One of METS's particular strengths is its ability to record complex links between different kinds of metadata and multiple digital objects. For instance, a METS wrapper can be used to encode and transmit all the metadata about individual page scans that make up a digital book. METS may be useful to the statistical community as a way to store and transmit metadata about complex data objects. METS is specifically designed to be encoded in XML and takes advantage of XML's hierarchical document structure.

ISO 19115

The ISO 19115: Geographic information metadata standard supports the documentation of geospatial data and services. This standard was harmonized from the Federal Geographic Data Committee's geospatial metadata standard and other commonly used geospatial standards in 2003. The resulting standard is widely adopted by geographic information system (GIS) software, and is important in facilitating the use, storage, management, and analysis of complex geographic datasets. These often include maps, charts, and text documents in addition to numeric and other non-geographic data. Many types of statistical data may have a geographic component or may need to be combined with geographic data. Additionally, some statistical metadata standards may be usable in combination with ISO 19115. Statistical agencies with a geographic focus should also review this standard.

PROV

The PROV family of standards was developed by the World Wide Web Consortium (W3C) to express provenance metadata for a wide range of information objects, including news articles, datasets, and more. Provenance metadata are important to scientific reproducibility and transparency, and

---

[25] See https://www.loc.gov/standards/mets/.

as such, could be important to include with statistical metadata. PROV models provenance as an interaction between three classes of elements: Agents, Entities, and Activities. Entities are the physical, digital, or conceptual things that are being created or altered; Agents are the people or software doing the alteration; and activities are the specific actions taken upon an Entity.[26] The W3C has developed specifications for encoding the PROV data model in a range of formats (e.g., XML and RDF), and for using it with a range of existing metadata standards (e.g., Dublin Core).

DCAT (Data Catalog Vocabulary)

A newer W3C standard, DCAT is the Data Catalog Vocabulary, a "vocabulary for publishing data catalogs on the web."[27] Where many of the standards being described in this document are meant to describe individual digital objects, DCAT is meant to describe the catalogs for those objects and therefore facilitate the sharing and aggregation of catalogs between/among multiple institutions or repositories. DCAT could therefore be useful to individual agencies or repositories wishing to eventually aggregate their data together.

Schema.org

Finally, schema.org is both the name and URL of a vocabulary created by Google, Microsoft, Yahoo, and Yandex, meant to semantically annotate resources on the Web. The schema.org terms describe a wide range of entities—everything from events to "friending actions" on social media to airline flight schedules—but most germane to this report is their set of terms for datasets and data catalogs. In short, this vocabulary is meant to be embedded in the HTML of the landing page of a dataset or catalog, where it can then be harvested by generic search engines such as Google. In this way, schema.org terms are quite different from traditional approaches to metadata, which are typically stored in a repository that cannot be Web indexed. A few scientific agencies are exploring the use of schema.org terms on their Web pages, and the use of Web protocols for dataset search (e.g., the EarthCube funded Project 418, https://github.com/earthcubearchitecture-project418). However, few repositories (if any) have yet to adopt these terms beyond pilot projects.

---

[26]See https://www.w3.org/TR/2013/NOTE-prov-primer-20130430/.
[27]https://www.w3.org/TR/vocab-dcat-2/#introduction.

Appendix B

The Role of Metadata in Assessing the Transparency of Official Statistics

INTRODUCTION

New expectations and requirements for finding, using, and understanding data are emerging rapidly, and the U.S. federal statistical community needs to keep pace to stay relevant. Transparency and reproducibility are two of these new general concerns, and this note is an attempt at laying out what they mean and how to achieve them. (In putting this together, we did not first perform a systematic literature review.)

As defined in dictionaries, transparency means "the condition of being transparent," and transparent means "easy to perceive or detect." So, for us, data or some other resources are transparent when it is easy to perceive or detect where they are, what they mean, and how to use them.

For our purposes, reproducibility is defined as "the extent to which consistent results are obtained when an experiment is repeated." The related term, repeatability, means "the closeness of the agreement between the results of successive measurements of the same measure carried out under the same conditions of measurement." Both these terms have a simpler interpretation in science than in federal statistics; however, they are useful for us as well. An example of reproducibility in official statistics would be to check that the same value of some measure (especially an economic indicator) remains the same after successive applications of the processing over the collected data. This becomes interesting when part of the process involves human judgment or some randomization. Replicability, on the other hand, would entail taking data from a different source representing the same reference period and checking to see if final results are consistent.

For instance, does some measure increase at the same rate as compared to one measured at an earlier time?

Both reproducibility and replicability depend on having access to and understanding of the data and the processing and analytical systems. This means descriptions of the data and the design, collection, processing, and estimation stages are accessible and understandable—in other words, transparent. Therefore, our focus will be transparency. This concept appears to be the more fundamental.

However, this general definition of transparency begs questions. How, in specific circumstances, do we ensure transparency? Is it possible to detect that some system is transparent? We know the answers will depend, in part, on the metadata that are available to a user. Metadata are the descriptive data for a resource, so the ability to find, understand, and use a resource depends on the quality of the metadata available. And here quality is a measure of how well the metadata provide all the information necessary to allow the user to complete the tasks set forth.

NATURE OF THE CONCEPT OF TRANSPARENCY

As defined above, transparency is a fairly general notion or concept. It is easy to exemplify, but difficult to define. By this, we mean that transparency is difficult to characterize in general and easier when applied to specific circumstances. We end up having to characterize transparency for each application, or kind of application. Why? Or better, what can we say to make this more tractable?

Cognitive psychologists have identified at least two kinds of concepts: entity concepts and relational concepts (Gentner and Kurtz, 2005). Entity concepts are easily characterized, so they can be directly measured. For example, *car* and *dataset* are entity concepts. Each specific one is easy to describe. Relational or role concepts are not so easily characterized. They are more general in nature and have to be refined, or specialized, in order to make them characterizable. An example is the word *guest*. One can be a guest at a party, a guest in a hotel, or a guest user on a secure Web site. These specific cases do not have much in common, and they are each characterizable.

In federal statistics, employment is an example of a relational concept. Employment by itself is too broad to characterize, but specific areas of employment can be characterized. Compensation, requirements, and employer costs, for example, are refinements that are more easily characterizable. Federal statistics have many such examples, and the concept of transparency has the same nature. In each situation where the issue of transparency comes up, it appears we need to characterize it in its own way.

METADATA SCHEMAS

Characterizing a concept provides a means to determine if a specific object or situation meets those criteria, that is, if the object or situation *corresponds* (as described further below) to the concept. We can define a tennis ball as a ball with certain specific features—pneumatic but not inflatable, specific colors, specific smell, fuzzy outside surface, 2.67 inches in diameter plus or minus some tolerance, weight of 2.04 ounces again subject to a tolerance, etc. So, if one is presented with a ball, it is pretty easy to determine if it is a tennis ball by examining the ball's properties.[1] That is, determining if a specific ball corresponds to being a tennis ball is easy.

The properties of a tennis ball, and those of any object in general, are descriptive of the object. Given the definition of metadata—data being used to describe some object(s)—the properties of a specific object are metadata.

The characteristics of a concept[2] are the categories to which the properties of objects that correspond to the concept belong. Tennis balls must weigh between 1.98 and 2.10 ounces. This is a characteristic. A particular tennis ball might weigh 2.01 ounces, and this is a property of that ball. Note, the weight of this ball (the property) corresponds to what a tennis ball needs to weigh (the characteristic).

The properties are metadata, and their corresponding characteristics are elements of a schema for that metadata. Each characteristic has a set of properties that correspond to it. For instance, the weight of a tennis ball can be any of the values between 1.98 and 2.10 ounces. These properties form the set of allowed values (or value domain) for the element resulting from the characteristic. So, if a particular concept is characterizable, these characteristics lead directly to a metadata schema. The elements and constraints of the schema turn the schema into a specification.

SPECIFICATIONS

A specification is a set of expressions called provisions. This is adapted from ISO/IEC Guide 2.[3] We define provision, and related terms, as follows:

---

[1] Properties differentiate objects from each other.
[2] Characteristics differentiate concepts from each other.
[3] https://ec.europa.eu/eurostat/cros/system/files/ISO%20reference%20definitions%20-%20%20guide%202%20-%202004%20-%20rev.doc.

- *provision*
  - expression in a specification that takes the form of a statement, an instruction, a recommendation, or a requirement[4]
- *statement*
  - expression that conveys information
- *instruction*
  - expression that conveys an action to be performed
- *recommendation*
  - expression that conveys advice or guidance
- *requirement*
  - expression that conveys criteria to be fulfilled.

For our tennis ball example, each characteristic or element is described in Table B-1, below. These descriptions contain provisions. For example, the optionality column determines whether an element is required (requirement), optional (recommendation), or conditional (instruction, "if label present"; and requirement). The conditional column contains instructions and requirements ("weight must be between 1.98 and 2.10 ounces").

The notion of conformance is used to determine and claim that a specification is being followed properly. Loosely speaking, conformance is the situation under which an object adheres to a specification. More precisely, conformance is defined as follows: fulfilment by a product, process, or service of specified requirements.

CONFORMANCE

It is easy to think that the only consideration in determining conformance is whether the requirement provisions are fulfilled, but that would be incomplete. In the tennis ball example below, another expression to determine whether a ball is a permissible tennis ball is to make sure the height of the bounce is between 53 and 58 inches after the ball is dropped onto a concrete floor from a height of 100 inches. Many of the provisions in that condition are instructions and statements, as we have seen.

So, through conformance claims it is possible to know formally if some specification is being followed. In the tennis ball example, we have a metadata schema in the form of a specification (ignoring color, smell, and fuzziness), as seen in Table B-1.

A ball is a permissible tennis ball if it satisfies all the requirements set forth in the schema above, i.e., the ball conforms to the specification. The

---

[4] These types of provisions are distinguished by the form of wording they employ; e.g., instructions are expressed in the imperative mood, recommendations by the use of the auxiliary "should," and requirements by the use of the auxiliary "shall."

TABLE B-1 Elements

Element name	Conditions	Optionality	Test or other criteria
Weight	1.98 < w < 2.10 oz	Required	
Diameter	2.57 < d < 2.70 in	Required	
Bounce	53 < b < 58 in	Required	Dropped 100 inches onto concrete
Label	Brand name	Optional	
Digit	Single-digit numeral	Conditional, if label present	Digit appears below the label

NOTE: This is not a complete list, and the optionality is modified for the purposes of illustration.

question of whether a tennis match is being played according to the rules comes down to whether the court, racquets, balls, and net conform to the specifications for each.

However, this schema provides multiple ways to conform. These options for conformance are known as varieties. In our example, the varieties arise, for instance, because of the "label" element. It is optional, meaning a ball can conform whether a label is printed on the ball or not. But, if there is a label on the ball, there must also be a digit printed on the ball as well. That digit must appear below the label. The solution is to state which optional elements are selected.

TRANSPARENCY

As the previous sections of this document describe, there are a series of considerations needed to ensure transparency in federal statistics. The important ones are these:

- Identify which aspect of the statistical business life cycle needs to be transparent to users;
- Identify the relevant information needed for the user;
- Either identify (from an existing specification or standard) or build the needed elements that can hold this information; and
- Create conformance criteria for this set of elements.

The considerations above embed the idea that transparency is very much a case-by-case problem. Since the sets of metadata elements needed to describe each part of the statistical business life cycle differ, sometimes markedly, the case-by-case approach is further justified.

The relationship between the information needed for transparency and the set of metadata elements in a specification supporting it is key. If the set of metadata elements is not complete in the sense of the information

TABLE B-2 Elements for Describing Variables

Element name	Conditions	Optionality	Test or other criteria
name	text	required	
universe	text	required	
question	text	required, if result of interview	result of answered question
derivation	formula	required, if calculated from other variables	formal language

needed, the required information will not all be available. We illustrate this with a simplified example in Table B-2 for describing a variable.

In the example in Table B-2, variables are described without enumerating or providing a rule for the allowed values. A simple example of an enumerated set of allowed values is {<M, male>, <F, female>} for a variable for "sex of a person." An example of a rule describing allowed values is {0 < = age < 100} for a variable "age in years of a person" (in this case, age is top-coded to 99).

Conforming to the specification laid out in Table B-2 does not make the information about variables transparent. There is too much missing information. The set of elements does not match all the information one needs to understand a variable.

One interesting addition to the story about conformance is the distinction between conformance and *strict* conformance. A system strictly conforms to a specification if it satisfies all the requirements and no others. Conformance, by itself, does not indicate whether this is the case.

A system conforming to the specification in Table B-2 could still be transparent if one extends the information provided by adding metadata elements. One could transform Table B-2 into Table B-3, with the extended element names in italics, where the last two rows are kinds of allowed

TABLE B-3 Extended Elements for Describing Variables

Element name	Conditions	Optionality	Test or other criteria
name	text	required	
universe	text	required	
question	text	required, if result of interview	result of answered question
derivation	formula	required, if calculated from other variables	formal language
allowed values	text	required	one of 2 kinds
<enumerated>		if applicable	set of ordered pairs <code, meaning>
<described>		if applicable	rule, formally written

values. Thus, the interpretation of the enumerated kind is further described in the second row and the described kind in the third row.

Now, Table B-3 provides transparency, provided the only necessary but missing information was the allowed values. Assuming this, all relevant information is provided in the metadata elements in the specification.

Appendix C

Public Meeting Agendas

MEETING 1
May 21, 2019
Keck Center of the National Academies, Room 100
500 Fifth Street NW, Washington, DC

Tuesday, May 21
OPEN SESSION

8:45 am   **A. Introduction of Panel and Staff**
Breakfast will be provided in Room 100

9:30   **B. Welcomes**
Emilda B. Rivers, *National Center for Science and Engineering Statistics (NCSES)*
Monica Feit, *Division of Behavioral and Social Sciences and Education*
Brian Harris-Kojetin, *Committee on National Statistics*

Questions from the Committee

10:15   *Break (informal discussion and further Q&A with presenters)*

10:30   **C. Charge from NCSES and the Federal Context**
         Overall Charge
                Emilda B. Rivers
         NCSES Data Management, Dissemination and User Perspective
                May H. Aydin, *NCSES*
         Open/Transparent Data Goals and Machine-Readable Data and
           Metadata
                Francisco Moris, *NCSES*
         Transparent Reporting for Integrated Data Quality: Assessing
           the User's Perspective
                Mark Prell, *Economic Research Service*
         Questions from Committee and Open Discussion

11:50   *Working Lunch in Atrium (continue discussions of presentations)*

12:50 pm **D. Challenges During the Statistical Survey Cycle:
         Data Documentation, Archiving, and Dissemination**
         Role of Contractors in NCSES Data Production, Archiving and
           Documentation
                May H. Aydin
         The Future of NCSES Data Systems: Metadata Explorer, APIs,
           and Interactive Data Tool
                Tiffany Ann Julian, *NCSES*
         Open Data and Metadata Schema
                Philip Ashlock, *General Services Administration*
         Documentation and Archiving Metadata Practices and Needs
                Marilyn Seastrom, *National Center for Education Statistics*

         Questions and Discussion

2:50   *Break (informal discussion and further Q&A with presenters)*

3:05   **E. Approaches to Data Documentation, Archiving, and
         Dissemination within the Federal Statistical System**
         Statistical Metadata and Statistics Registries
                Daniel W. Gillman, *U.S. Bureau of Labor Statistics*
         Statistical Workflow Processes and Internal Metadata Capture/
           Management
                Christopher N. Carrino, *U.S. Census Bureau*
         Transparency, Quality, Documentation, and Dissemination:
           International Standards
                John L. Czajka and Mathew Stange, *Mathematica Policy
                Research*

Questions and Discussion

4:50 Concluding Remarks
Emilda B. Rivers, May H. Aydin, Francisco Moris, and Michael Cohen

5:05 pm *Adjournment*

MEETING 2
September 9, 2019
Keck Center of the National Academies
500 Fifth Street NW, Washington, DC

Monday, September 9, 2019, Room 100
OPEN SESSION

8:45 am Welcome/Introductions/Goals for the Day
Breakfast will be provided outside of Room 100

9:00 Driving Official Statistics to its Modernization
Juan Muñoz, *National Institute of Statistics and Geography (Mexico)*

9:45 Producing the National Income and Product Accounts: An Illustration Using Personal Consumption Expenditures
Dennis Fixler, *Bureau of Economic Analysis*

10:30 *Break (informal discussion and further Q&A with presenters)*

10:45 Transparency and Reproducibility in the Design, Testing, Implementation and Maintenance of Procedures for the Integration of Multiple Data Sources
John Eltinge, *U.S. Census Bureau*

11:30 Data Curation and Transparent Federal Statistics: Some Suggestions
Leighton L. Christiansen, *Bureau of Transportation Statistics*

12:30 pm *Working Lunch (continue discussions of presentations)*

1:30 Metadata Driven Statistical Data Management
Pascal Heus, *Metadata Technology North America*

2:30	**Building an Open Ecosystem for Data Discovery** Natasha Noy, *Google*
3:30	*Break (informal discussion and further Q&A with presenters)*
3:45	**Transparency for the Modern Evidence Ecosystem: Trust and Accountability to Support Effective Policymaking** Nick Hart, *Data Coalition*
4:30	**Recent OMB Efforts to Increase the Transparency and Reproducibility of Federal Statistics** Nancy Potok, *Office of Management and Budget*
5:00	**Closing Remarks** Staff from *National Center for Science and Engineering Statistics*
5:15 pm	**Adjourn Open Session**

MEETING 3
November 5–6, 2019
Keck Center of the National Academies
500 Fifth Street NW, Washington, DC

Tuesday, November 5, 2019
OPEN SESSION, Room 208

8:45 am	**Welcome/Introductions/Goals for the Day** Dan Kasprzyk, *Panel Chair* Emilda Rivers, *National Center for Science and Engineering Statistics*
9:00	**Overview and Uses of SDMX** David Barraclough, *OECD*
10:00	*Break (informal discussion and further Q&A with presenters)*
10:15	**Census Bureau's Approach to Data Dissemination** Zach Whitman, *U.S. Census Bureau*
11:15	**Disseminating Methodological Processes** Bill Bell, *U.S. Census Bureau*

APPENDIX C 231

11:45	*Working Lunch (continue discussions of presentations)*
1:00 pm	**Transparency and User's Needs** Olivier Dupriez, *World Bank*
1:45	**Uses of Various Metadata Standards and Tools Outside of U.S.** Jeremy Iverson and Dan Smith, *Colectica* Heidi Koumarianos, *National Institute of Statistics and Economics Studies (France)*
3:15	*Break (informal discussion and further Q&A with presenters)*
3:30	**Role of Contractors in Archiving Data and Methods** Brad Edwards, *Westat* Marcus Berzofsky, *RTI International* Marilyn Seastrom, *National Center for Education Statistics*
4:45	**Closing Remarks** Staff from *NCSES*
5:00 pm	*Adjourn Open Session*

MEETING 4
February 6, 2020
Keck Center of the National Academies
500 Fifth Street NW, Washington, DC

Wednesday, February 6, 2020
OPEN SESSION, Room 106

8:45 am	**Welcome/Introductions/Goals for the Day** *(Breakfast available at 8:30 am)* Dan Kasprzyk, *Panel Chair* Emilda Rivers, *NCSES*
9:00	**Transparency and User's Needs** Olivier Dupriez, *World Bank*
10:00	*Break (informal discussion and further Q&A with presenters)*
10:15	**Experiences in Analyzing NCSES Data** Anne-Marie Knott, *Washington University in St. Louis* Kimberlee Eberle-Sudre, *Association of American Universities*

11:00 **American Community Survey Data Users Group and Related Activities**
Jason Jurjevich, *University of Arizona*
Tori Velkoff, *U.S. Census Bureau*

11:45 am ***Adjourn Open Session***

Appendix D

Biographical Sketches of Panel Members

DANIEL KASPRZYK (*Chair*) is a consultant and senior fellow at NORC at the University of Chicago. Prior to his appointment at NORC, he was vice president and managing director of surveys and statistics at Mathematica Policy Research, Inc. Kasprzyk has more than 30 years of experience in managing large-scale sample surveys in a variety of topic areas, including holding various positions on the staff of the Survey of Income and Program Participation at the U.S. Census Bureau and carrying out methodological research associated with federal survey programs. He has particular expertise in nonsampling error issues in surveys. Prior to his private-sector positions, he was program director of the Elementary and Secondary Sample Survey Studies Program at the National Center for Education Statistics, where he was responsible for the Schools and Staffing Survey System. He was a member of the Organisation for Economic Co-operation and Development (OECD) committee that developed and reported school and teacher data for national comparisons. He served as the U.S. Department of Education's liaison to the National Academy of Sciences' Panel on Estimates of Poverty for Small Geographic Areas. He was a member of the National Academy of Sciences' Panel to Review the 2014 Redesign of the Survey of Income and Program Participation, a member of the National Academy of Sciences' Panel to Review the National Children's Study Research Plan, and a member of the Institute of Medicine's Panel on Lesbian, Gay, Bisexual, and Transgender Health Issues and Research Gaps and Opportunities. He also served for 20 years on the Office of Management and Budget's Federal Committee on Statistical Methodology and chaired committees on federal longitudinal surveys and data quality reporting and measurement at the Center for Excellence in Survey Research at NORC at the University of Chicago. He is an elected member of the International Statistical Institute and fellow and former vice president of the American Statistical Association (ASA). He chaired the ASA Sections on Survey Research Methods and on Social Statistics, as well as serving as officer for other sections of the ASA and for the Washington Statistical Society, a chapter of the ASA. Kaspryzk has a B.S. in mathematics from Wayne State University and a Ph.D. in mathematical statistics from George Washington University.

PHILIP ASHLOCK leads the data and analytics portfolio at the GSA Technology Transformation Service and serves as the chief architect for Data.gov. At Data.gov, he oversees an open development process and a federated architecture supporting open data and application programming interfaces (APIs) across government. Recently, Ashlock launched the U.S. Data Federation, a new project exploring reusable tools and repeatable processes to support data standards and interoperability within government. Previously, he served as a presidential innovation fellow working with the GSA and the White House Office of Digital Strategy. In addition to overseeing metadata management and standards across both federal agencies and local government in the United States, he has actively participated in international efforts around data standards and open data. This includes work on the Open311 API standard for municipal service requests, the international standards for data catalog metadata like DCAT, support on the U.S. National Reporting Platform for the Sustainable Development Goals, and participation in the UN SDMX-SDGs Working Group. Ashlock has a B.A. in design with a concentration in new media and a computer science minor from Western Washington University.

DAVID BARRACLOUGH is a 29-year veteran of the information technology (IT) industry, the past 15 years working in official statistics on IT systems, architecture, and methodology. He is currently a smart data practices manager in the OECD Statistics and Data Directorate, where he leads a team and community that is remodeling all of the OECD's disseminated data using the SDMX standard, and designing the accompanying methodology. He is chair of the SDMX statistical working group (held for 5 years), which maintains the Content-Oriented Guidelines, including how to implement SDMX, cross-domain concepts code lists, how to model statistical datasets, and many other instruments and guidelines. He also manages the SDMX for Labor Statistics Global Data Structure Definition project—the first Global DSD for social statistics, and is involved in other domains such as the Sustainable Development Goals, Education, and National Accounts. He is also involved with statistical modernization standards such as GSBPM, GSIM, and CSPA. Barraclough attended Barnsley College in the UK.

CHRISTOPHER CHAPMAN joined the National Center for Education Statistics (NCES), within the U.S. Department of Education's Institute of Education Sciences, in 1997. Since joining NCES, he has held a number of positions, starting as project officer for household surveys conducted by NCES. During his career at NCES, he has led or contributed significant methodological guidance to dozens of large-scale sample surveys such as the early childhood longitudinal studies, surveys of school safety, recurring teacher surveys, and the American Community Survey. He is currently associate commissioner of NCES for their Sample Surveys Division. In addition to his work with NCES, Chapman is a member of the interagency Federal Committee on Statistical Methodology, where he is currently coauthoring a report on how agencies can best communicate complexities of mixed-source data products to the wide range of stakeholders who use and rely on federal data. Prior to joining the Department of Education, he worked at the American Institutes for Research (AIR) where he was the project lead on work with NCES to strengthen and improve its household surveys. Before AIR, he worked at the Ohio State University's Center for Survey Research (then the Polymetrics Laboratory), where he collaborated with academic researchers, state and local government officials, and private firms to develop and field a large number of data collections. Chapman has a B.A. and an M.A. in political science from the Ohio State University.

DANIEL W. GILLMAN is a mathematical statistician at the U.S. Bureau of Labor Statistics (BLS) in the Office of Survey Methods Research. His research interests include metadata, standards, terminology, and classification. At BLS, he led the effort to build a taxonomy of terms describing all time-series data and was a member of the team to build a glossary of BLS technical terms. He is consultant to the Consumer Expenditure Surveys effort to build a metadata repository in support of the annual public-use microdata release, to the BLS output database redesign effort, and to data-governance modernization efforts at the U.S. Department of Labor. He is chair of the interagency SCOPE/Metadata interest group to develop guidance on metadata management for the statistical agencies. Previously, Gillman chaired the Federal Data Architecture Subcommittee/Open Government Vocabulary Working Group, the Statistical Data and Metadata Exchange/Statistical Working Group, and the International Committee for Information Technology Standards/Metadata Standards Technical Committee (L8). He is a member representative to the Data Documentation Initiative (DDI) Alliance and is a key developer of the DDI-4 (Cross-Domain Integration) model-driven standard. Under the United Nations Economic Commission for Europe (UNECE), he was a member and chair for the former Statistical Metadata Working Group, a key developer of several UNECE statistical metadata standards, and a member of the Supporting Standards Group.

Prior to working at BLS, he worked at the U.S. Census Bureau. Gillman has a B.S. and an M.A. in mathematics from the University of Maryland.

LINDA A. JACOBSEN is vice president of U.S. Programs at the Population Reference Bureau (PRB). She is a demographer with more than 30 years of experience analyzing population trends and their implications for professional, policy, and media audiences. Her research has focused on family and household change, child and family well-being, and population estimates and projections. In partnership with the U.S. Census Bureau, Jacobsen leads several projects to increase knowledge and use of the American Community Survey (ACS) and to collect data-user feedback on ACS and decennial census products. She also directs PRB's Center for Public Information on Population Research, funded by the Eunice Kennedy Shriver National Institute of Child Health and Human Development. Jacobsen has been a featured speaker on U.S. demographic trends at Harvard University's Program for Newly Elected Members of Congress, the Knight Center for Specialized Journalism, and many other professional meetings and conferences. She has served on the Census Bureau's Scientific Advisory Committee, a National Academy of Sciences' Panel on the ACS, and as chair of the Population Association of America (PAA) Committee on Government and Public Affairs. She currently chairs the board of directors of the Council of Professional Associations on Federal Statistics and serves on PAA's Committee on Population Statistics. She was elected a fellow of the American Statistical Association in 2015. Before joining PRB in 2005, she served as a senior executive and chief demographer for two leading marketing information companies; the research director at *American Demographics*; and a faculty member at both Cornell University and the University of Iowa, where she conducted research and taught graduate studies in sociology and demography. Jacobsen holds an M.S. and a Ph.D. in sociology from the University of Wisconsin–Madison and a bachelor's degree in sociology from Reed College.

H.V. JAGADISH is Bernard A. Galler collegiate professor of electrical engineering and computer science in the Department of Computational Medicine and Bioinformatics at the University of Michigan and director of the Michigan Institute for Data Science. His research focuses on how to build database systems and query models so that they are truly usable, and how to design analytics processes so that they can deliver real insights to nontechnical decision makers. His current research is centered on usability of Big Data, particularly when the data involved come from multiple heterogeneous sources, and have undergone many manipulations. He is an elected fellow of the Association for Computing Machinery and serves on the board of the Computing Research Association. Jagadish has a Ph.D. in electrical engineering from Stanford University.

FRAUKE KREUTER is director of the Joint Program in Survey Methodology (JPSM) at the University of Maryland, College Park; professor of statistics and methodology at the University of Mannheim; and head of the Statistical Methods Research Department at the Institute for Employment Research in Nürnberg, Germany. Before joining the University of Maryland, she was a postdoc at the University of California, Los Angeles, Statistics Department. Her research focuses on sampling and measurement errors in complex surveys. In her work at JPSM, she maintains strong ties to the federal statistical system, and serves in advisor roles for the National Center for Education Statistics and the U.S. Bureau of Labor Statistics. She has served as a member on the Panel on Improving Federal Statistics for Policy and Social Science Research Using Multiple Data Sources and State-of-the-Art Estimation Methods at the National Academies of Sciences, Engineering, and Medicine. She is the author or coauthor of several books, including *Data Analysis Using Stata* and *Practical Tools for Designing and Weighting Survey Samples*. Kreuter has an M.A. in sociology from the University of Mannheim, Germany, and a Ph.D. in survey methodology from the University of Konstanz.

MARGARET LEVENSTEIN is director of the Inter-university Consortium for Political and Social Research at the University of Michigan, research professor for both the Survey Research Center and the School of Information, and adjunct professor of business economics and public policy at the Ross School of Business at the University of Michigan. Levenstein first joined the Institute for Social Research's (ISR's) Survey Research Center in 2003 as executive director of the Michigan Census Research Data Center, a joint project with the U.S. Census Bureau. She has taken an active role at ISR, joining the director's advisory committee on diversity in 2009, and serving as chair of ISR's Diversity, Equity, and Inclusion Strategic Planning Committee, and as the liaison to the larger university program. Her research and teaching interests include industrial organization, competition policy, business history, data confidentiality protection, and the improvement of economic statistics. She is associate chair of the American Economic Association's Committee on the Status of Women in the Economics Profession and past president of the Business History Conference. Levenstein has a B.A. from Barnard College, Columbia University, and a Ph.D. in economics from Yale University.

PETER V. MILLER is a retired senior researcher for survey measurement at the U.S. Census Bureau. He joined the staff of the Census Bureau as chief of the Center for Survey Measurement in 2011. He was named chief scientist of the Bureau's Center for Adaptive Design in 2013. Prior to joining the Census Bureau, he served on the faculty of Northwestern University for 29 years. He also held faculty positions at the University of Michigan, the University of Illinois, and Purdue University. While in federal service, he served as a member of the Federal Committee on Statistical Methodology (FCSM). He co-chaired the FCSM nonresponse bias working group and the adaptive design interest group. He also co-chaired a task force on improving the climate for surveys, sponsored by the American Association for Public Opinion Research (AAPOR) and the American Statistical Association (ASA). Miller has held several elective offices in AAPOR, serving as president from 2009 to 2010. During his tenure as president, he launched the association's Transparency Initiative. His research interests are centered in survey data collection methodology and transparency policies and procedures. He was named a fellow of the ASA in 2015. Miller has an A.B. and a Ph.D. both from the University of Michigan.

AUDRIS MOCKUS is Ericsson-Harlan D. Mills chaired professor of digital archeology in the Department of Electrical Engineering and Computer Science at the University of Tennessee, Knoxville. He also works part time at Avaya Labs Research. Mockus studies software developers' culture and behavior through the recovery, documentation, and analysis of digital remains. These digital traces reflect projections of collective and individual activity. He reconstructs the reality from these projections by designing data mining methods to summarize and augment these digital traces, interactive visualization techniques to inspect, present, and control the behavior of teams and individuals, and statistical models and optimization techniques to understand the nature of individual and collective behavior. He is a member of the Institute of Electrical and Electronics Engineers and the Association for Computing Machinery. He served on the National Academies of Sciences, Engineering, and Medicine's Steering Committee for Transparency and Reproducibility in Federal Statistics: A Workshop. Mockus has a B.S. and an M.S. in applied mathematics from the Moscow Institute of Physics and Technology, and a Ph.D. in statistics from Carnegie Mellon University.

SARAH M. NUSSER is professor of statistics at Iowa State University and affiliated with its Center for Survey Statistics and Methodology (CSSM), which she directed for 15 years. She is visiting professor at University of Virginia's Social and Decision Analytics Division of the Biocomplexity Institute and senior fellow with the Association of American Universities (AAU). She previously served as Iowa State University vice president for research for 6 years. Nusser's current research focuses on improving the reusability and impact of publicly accessible research data. Her prior research focused on survey statistics and methodology for land-based and population-based surveys, including sampling and estimation for longitudinal natural resource surveys, measurement error models for dietary intake and physical activity data, and geospatial methods for sample frame listing and natural resource surveys. She directed statistics and methodology research and development for the annual U.S. Department of Agriculture's National Resources Inventory Program for 22 years through her affiliation with CSSM. Nusser is actively involved in efforts to promote open science, transparency, and public access to research data. She serves as chair of the National Academies of Sciences, Engineering, and Medicine's Board on Research Data and Information and is a former member of the Committee on National Statistics. She is a member of the National Institutes of Health's Director Advisory Committee Working Group on Enhancing Rigor, Transparency and Translatability in Animal Research. She has played leadership roles in the AAU's Association of Public and Land-grant Universities initiative on Accelerating Public Access to Research Data since its inception in 2017. Nusser is fellow of the American Statistical Association and elected member of the International Statistical Institute and has served on numerous scientific panels, advisory committees and governing boards. Nusser has a B.S. in botany from the University of Wisconsin, an M.S. in botany from North Carolina State University, and a Ph.D. in statistics from Iowa State University.

ERIC RANCOURT is director general of the Modern Statistical Methods and Data Science Branch at Statistics Canada, where he has been for 30 years. He has occupied several roles, such as director general of strategic data management, director of international cooperation, director of corporate planning, head of research, production manager of *Survey Methodology Journal*, and researcher. His main areas of work have been on treatment of nonresponse, estimation, gathering, safeguarding, and use of administrative and alternate data in statistical programs. He has been involved in many professional associations and is an International Statistical Institute elected member. He has a B.A. in statistics from the Université Laval.

LARS VILHUBER is on the faculty of the Department of Economics at Cornell University, a senior research associate at the Industrial and Labor Relations (ILR) School at Cornell University, and executive director of ILR's Labor Dynamics Institute. He is also senior research associate (IPA) at the Center for Economic Studies and LEHD Program at the U.S. Census Bureau. Vilhuber has worked in both research and government and he has consulted with government and statistical agencies in Canada and the United States. Currently, he conducts research on using and making available highly detailed longitudinally linked data to analyze the effects and causes of mass layoffs, worker mobility, and the dynamics of (local) labor markets. These data are generally subject to severe access restrictions. In order to make such data available to other researchers, he also conducts research on statistical disclosure limitation issues, including the creation and dissemination of synthetic data, and investigates novel methods and tools to disseminate metadata on such data. He is currently principal investigator on numerous grants, including those that fund activities at the National Science Foundation-Census Research Network (NCRN) node at Cornell University. He is the lead principal investigator on the NCRN Coordinating Office. Vilhuber has an undergraduate degree in economics from Universität Bonn, Germany, and a Ph.D. in economics from Université de Montréal, Canada.

COMMITTEE ON NATIONAL STATISTICS

The Committee on National Statistics was established in 1972 at the National Academies of Sciences, Engineering, and Medicine to improve the statistical methods and information on which public policy decisions are based. The committee carries out studies, workshops, and other activities to foster better measures and fuller understanding of the economy, the environment, public health, crime, education, immigration, poverty, welfare, and other public policy issues. It also evaluates ongoing statistical programs and tracks the statistical policy and coordinating activities of the federal government, serving a unique role at the intersection of statistics and public policy. The committee's work is supported by a consortium of federal agencies through a National Science Foundation grant, a National Agricultural Statistics Service cooperative agreement, and several individual contracts.